ARMORED CAV

A Guided Tour of an Armored Cavalry Regiment

Tom Clancy's
Bestselling Novels Include:

The Hunt for Red October

Red Storm Rising

Patriot Games

The Cardinal of the Kremlin

Clear and Present Danger

The Sum of All Fears

Without Remorse

Debt of Honor

Nonfiction:

Submarine:
A Guided Tour Inside a Nuclear Warship

Armored Cav:
A Guided Tour of an Armored Cavalry Regiment

TOM CLANCY

ARMORED CAV

*A Guided Tour
of an Armored Cavalry
Regiment*

B

BERKLEY BOOKS, NEW YORK

The views and opinions expressed in this book are entirely those of the author and do not necessarily correspond with those of any corporation, navy, or government organization of any country.

ARMORED CAV

A Berkley Book / published by arrangement with
Jack Ryan Limited Partnership

PRINTING HISTORY
Berkley trade paperback edition / November 1994

ISBN: 0-425-15836-5

BERKLEY®
Berkley Books are published by
The Berkley Publishing Group, 200 Madison Avenue,
New York, New York 10016.
BERKLEY and the "B" design are trademarks of
Berkley Publishing Corporation.

PRINTED IN THE UNITED STATES OF AMERICA

10   9   8   7   6   5   4   3   2   1

This book is dedicated to the troopers of the 11th and 14th Armored Cavalry Regiments. As the last of them stand down from their almost five-decade vigil over the Fulda Gap in Germany, they can take pride that they won their war without a shot having to be fired in anger. May they find, in life and beyond, the peace that they spent their lives forging and protecting for the rest of us. God bless, guys.

The author gratefully acknowledges permission for use of the following photos and other materials:

The United States Army; H. R. McMaster; Toby Martinez; AM General Corporation; BEI Defense Systems Company; Bell Helicopter-Textron, Inc.; Beretta USA Company; BMY; Boeing-Sikorsky; Colt's Manufacturing Company; FMC Corporation; General Dynamics Land Systems; Loral Vought Systems; McDonnell Douglas Helicopter Company; Oshkosh Truck Corporation; Sikorsky Aircraft; Trimble Navigation Ltd.; Hughes Missile Systems Company; John D. Gresham; all rights reserved.

Contents

Acknowledgments

O nce again it is time to say thanks to all the people who really made this book a work to be proud of. At the start is my partner and researcher, John D. Gresham. Once again, he has literally traveled from coast to coast to assemble the wealth of material and experiences that are being presented for your enjoyment and enlightenment. His is a huge job with some deeply difficult tasks associated with it, and my best thanks go to him for his continued dedication and friendship. Also, the entire team has again benefited from the wise counsel and advice of series editor Professor Martin H. Greenberg. Laura Alpher is to be complimented again for her marvelous drawings which have added so much to the quality of the final book. This young lady is a major talent, so keep your eyes out for her work in the future. Tony Koltz, Mike Markowitz, and Chris Carlson also need to be recognized for their research and editorial support that was both critical and timely. A special note of appreciation goes to Greg Stewart for his fine photographic advice. Thanks also to Cindi Woodrum, Diana Patin, and Roselind Greenberg for their patience and support in backing the rest of us up as we went along digging out the bits and pieces of our little story.

One thing that makes a book both fun and exciting to work on is support from on high, and this book was blessed with Army support that was, in a word, unbelievable. It is difficult to say thanks enough to the U.S. Army Chief of Staff, General Gordon Sullivan, U.S.A. This gentleman soldier, who looks like a kindly pet shop owner, is the driving force behind the technical revolutions that are currently being implemented in the Army. The Army and the country are blessed to have his leadership at this critical time in the history of our army. Also deserving special thanks is General Fred Franks, U.S.A., for his time and patience, as well as his friendship to all the members of our team. Thanks also to General Barry McCaffrey, U.S.A., and his wife, Jill, for sharing a special evening at their home with us. Down at Fort Hood, Texas, our appreciation goes out to my old friend, Lieutenant General Pete Taylor, U.S.A., and his replacement, Lieutenant General "Butch" Funk, U.S.A. Out at the National Training Center at Fort Irwin, California, there was Brigadier General Bob Coffey, U.S.A., who took time out of his busy schedule to show us the world's finest ground warfare training center. Thanks also to Brigadier General Harold Wilson, U.S.A., Ph.D., the chief historian for the U.S. Army.

Finally, there were three extraordinary young officers: Captains H. R. McMaster and Joseph Sartiano, and 1st Lieutenant Dan Miller, who shared their own Gulf War experiences with us. I think you will be as amazed as we were when you read them.

Another group that was vital to our efforts, less well known but equally important, were the members of the various U.S. Army public affairs offices (PAOs) and protocol organizations that handled our numerous requests for visits and information. Tops on our list was Major Rick Thomas of the Pentagon PAO. Rick and his team helped grant virtually every wish for information and access that we had, and made the whole security review process a pleasure. Over in General Sullivan's office, Lieutenant Colonel Robert Coffey, U.S.A., helped lay out the Army's modernization plans. Out at Fort Irwin, California, Captains Franklin Childress and Len Tokar made our visit both memorable and livable in the incredible heat of September, 1993. Down at Fort Polk, Louisiana, Lieutenant Colonel Mike Trahan, U.S.A., as well as Dan Nance and Dave Bingham helped get the light cavalry story across to us. Down at TRADOC Headquarters, Colonel George Stinnett and Ray Harper worked miracles to support our efforts.

Down at Fort Bliss, Texas, we had the honor of meeting as fine a group of warriors as you can imagine in the troopers of the 3rd Armored Cavalry Regiment. Our first thanks go to the regimental commander, Colonel Robert Young, U.S.A. This lifelong cavalry trooper, who doubles as a peacekeeper and relief worker, is a man of amazing contrasts. Our thanks for sharing time out of the regiment's training schedule to school us on the ways of the 3rd's troopers. In addition, the regimental staff deserves some mention here. Command Sergeant Major Dennis E. Webster and the regimental executive officer, Lieutenant Colonel Luke Barnett, U.S.A., did a fine job of coordinating our visits to the regiment. And the regimental PAOs Captain Andy Vliet and 2nd Lieutenant Nichole Whitehead were fantastic in their tolerance and patience. We also want to recognize the assistance of the regiment's various squadron commanders: Lieutenant Colonels Norman Greczyn, Karl J. Gunzelman, Gratton Sealock, and Thomas M. Hill. And of course, there were the extraordinary efforts of Lieutenant Colonel Toby W. Martinez, the commander of the regiment's 1st (Tiger) Squadron. Toby is one of the best cavalry officers in the Army today. He tolerated having us there to watch his victories, his defeats, and his learning experiences. God bless, Toby.

Many thanks are due to our various industrial partners, without whom all the information on the various weapons and systems would never have happened. At the armor manufacturers there was Carl Oskoian of General Dynamics, Ken Julian and Judy McIlvanie of BMY, as well as Bill Highlander and Madeleine Orr-Geiser of FMC. Among those at the helicopter builders were Russ Rumney at Bell, Jim Kagdis at Boeing-Sikorsky, Ken Jensen at McDonnell Douglas, and finally, Bill Tuttle and Foster

Morgan of Sikorsky. The folks at the truck contractors were a wealth of data, with Walt Garlow and Lynn Jones at Oshkosh Truck, as well as the incomparable Craig MacNab at AM General, as standouts. We made many friends at the various missile, armament, and system manufacturers including: Natalie Riley at BEI, Russ Logan at Beretta USA, Cynthia Pulham at Boeing, Art Dalton and Brian Berger at Colt, Clementine Cacciacarro and Cheryl Wiencek at Hughes, Robert Clower at IDA, Tommy Wilson at Loral, Jody Wilson-Eudy at Motorola, Ed Alber at Olin, Jim Walker at Rockwell, Peter Jones of Tenebrex, and last, but certainly not least, Ed Rodemsky and Barbara Thomas of Trimble, who spent so much time and capital to educate us on the subtleties of the GPS system. Thanks to you all.

Once again, our thanks for all of our help up in New York. To Robert Gottlieb at William Morris, thanks again for a great opportunity. And at Berkley Books, our appreciation goes out to our editor, John Talbot, as well as to Jacky Sach, Patty Benford, and David Shanks. To our friends like Ed Burba and Donn Starry, thanks again for your contributions and wisdom. To all the guys who took us for rides, thanks for teaching the heathens how things really work. For our families and friends, we once again thank you for having endured late dinners, broken promises, and missed holidays; please know that we love you always. And lastly, to the diminutive tank sergeant known as "Big Daddy," we say "...Thanks for going armor!"

Foreword

"IF YOU AIN'T CAV, YOU AIN'T..." Frequently displayed in conspicuous places, that trenchant soldier slogan reflects the cavalry trooper's perception of himself and his outfit. This splendid book is about those soldiers. Specifically it is about armored cavalry and about one armored cavalry regiment, the 3rd (known as the "Brave Rifles"), soon to be the last such organization on the active rolls of the U.S. Army—an army which in the not-too-distant past boasted five such active regiments, and among whose most effective unit performances in the Vietnam War were those of an armored cavalry regiment, the 11th (known as "The Blackhorse"). An army whose deployed armored cavalry regiments, the 2nd ("Toujours Prêt") and the 3rd, were among the most effective in the Gulf War, and in which there is a rich tradition of outstanding cavalry units and famous senior leaders who served as younger officers in those units. In this book Tom Clancy describes better than anyone yet why cavalry is what it is—why it is different—and why the nation is likely to need more, not fewer, cavalry regiments as we probe tentatively toward the 21st century.

Like all military units, modern armored cavalry is a unique combination of soldiers and equipment—but most of all, it's the soldiers who are important. A persistent lesson of battle is that, however good the equipment, what wins an engagement is the combination of the courage of the soldiers, the excellence of the leadership, and the effectiveness of the training—individual and small-unit training in employment of the equipment according to well-thought-out and frequently practiced tactics and operational schemes. So it all goes together—tactics, equipment, training, and organization. All that is what this book is about: a special organization, its equipment, its soldiers, its leaders, and how it fights. It's not just in cavalry units that all that goes together, for it does so in many units. But several characteristics of cavalry make that organization perhaps more *"different"* than other organizations. What are those characteristics?

First are the traditional cavalry missions—reconnaissance and security. In other words you have to find an enemy force in order to inform a higher command; then you may have to take action to prevent that enemy force from interfering with friendly force battle plans. These historic cavalry missions demand great operational mobility, tactical agility, superb command and control, and a special ability to operate effectively over vast distances. Additionally, they require the ability to concentrate quickly to meet a threat or to take advantage of an opportunity. In sum, they require an ever-present need to seize the initiative. The history of battle teaches the importance of seizing the initiative. For whoever takes the initiative usually wins—regardless of who may outnumber or be outnumbered, or who attacks or defends. Cavalry missions require units organized and trained to take the initiative, all the time, from the very beginning, every time out.

Second, cavalry organization *is* quite different. The need for mobility and agility, and for economy of force over vast distances, has caused modern cavalry to be organized around combined-arms teams at the lowest levels of command. Postwar armored cavalry units from the platoon level up consisted of a built-in mixture of scouts, infantry, tanks, and mortars for indirect fire support. Thus, an armored cavalry second lieutenant commands a full spectrum of combined-arms capabilities in his own little piece of the Army. With several such platoons, the cavalry troop commander can use the platoons as is, or he can group tanks, infantry, mortars, and scouts from all platoons in a combined-arms team at the troop level. In squadrons of armored cavalry regiments, a tank company and a self-propelled artillery battery give squadron commanders yet a stronger combined-arms team. With three cavalry squadrons, an air cavalry squadron of scout and attack helicopters, combat engineer and chemical companies, and a combat support squadron providing organic logistics support, the regimental commander has a truly imposing independent combined-arms force. An armored division several times the size of such a regiment would be required to provide an equal capability. As we shall see in Tom Clancy's account, with the advent of more complex weapons systems, levels at which combined arms are grouped have moved up from platoon to troop. But the result is the same—from their first day of duty cavalry leaders think *combined-arms warfare.* And that is a major reason why they are different, for battle experience teaches that the most difficult task to learn is the art—and it is an art—of knowing instinctively how to employ combined arms. Cavalry leaders are forced to do just that every day of their lives.

In Tom Clancy's book we meet two Persian Gulf War cavalry leaders who are masters of that art—one an armored corps commander, the other a cavalry troop commander. We learn that the corps commander, General Fred Franks, was once himself a cavalry platoon leader, later a squadron commander, and still later commander of a regiment, the 11th. The cavalry troop commander, Captain H. R. McMaster, Jr., began his own career as a second lieutenant in a cavalry unit. Each, at his own level, is a master of the art of combined-arms warfare.

The outstanding battle record of armored cavalry in the Gulf War is an important part of Tom Clancy's account, especially in the interviews with General Franks and Captain McMaster. It is well to remember the equally outstanding battle performance of armored cavalry in the Vietnam War, for there it was originally perceived that the nature of the war, the enemy, terrain, and climate would make it impossible to employ *any* armored units. Consequently, infantry divisions initially deployed to Vietnam without organic tank battalions and armored cavalry squadrons. But after some experience, the infantry commanders already in-country sent home for those units. With some misgivings, a regiment of cavalry, the 11th, was deployed to Southeast Asia. The misgivings proved unfounded. The 11th did very well. Other units were sent, and they also performed well. Later, a special task force sent to evaluate the performance of armored units in Vietnam concluded they were the most effective, certainly the most cost-effective, of the war. And when the war ended, cavalry units were among the last to redeploy home; they simply represented more combat power for the least investment in manpower that could be had.

As a result of this experience, it was quite clear that cavalry was a force ready-made for the less-than-all-out wars that became the norm at the end of the Cold War and during the years that followed. Changes to organization and equipment suggested by cavalry commanders were the basis for post-Vietnam cavalry reforms, and for the success of splendid regiments like the 3rd that were so effective in the 1991 Persian Gulf War.

Now our once-familiar world has changed. The traditional threat of more than five decades' standing has come asunder. And no one can reasonably say what will eventuate. A Russia there will be, but what kind? And what may be its relationship(s) with the other former Soviet republics? Four of those republics are now proud owners of thermonuclear weapons and the means to deliver them inter- or intra-theater. Meanwhile, militarization of conflict in the Third World continues apace. Indeed it is at a new juncture, with the advent of ballistic missiles and the drive by rogue dictators in resource-critical regions to develop or acquire weapons of mass destruction. Economic interdependence, an inescapable fact of our "modern" world, makes all nations, especially developed nations, vulnerable to mischief by Third World miscreants. At the same time, domestic economic uncertainty and national priorities now aimed at medical, educational, urban, and environmental dilemmas make it difficult to garner support for military forces to counter these new and unfamiliar threats.

The United States historically follows successful military ventures by destroying the very strength that provided victory—only to find it necessary to rebuild on short order with the arrival of new and previously unforeseen threats. The first battles of the next war thus find us paying the tragic price for our unpreparedness in the commodity we can least afford to expend—the lives of soldiers, marines, airmen, and sailors. Now we are at it once more. The situation of this last decade of the 20th century is reminiscent of the thirty years following the American Civil War, when shadowed remnants of a once-proud army were completely overcommitted in campaigns to tame the American West and its Indians. Now, the "wild savages" of the American West have been replaced by religious fundamentalists and other types of "hostiles." Though they are geographically far away, their weapons put them relatively closer than those now all-but-forgotten frontier "badmen." Yet the drive to disarm continues. The American Congress is frequently blamed for this state of affairs, and not without justification. For the Congress, for better or for worse, has to determine the size of the armed forces in budget terms—numbers of people, of bases, and of kinds of equipment. On the other hand, structure—the numbers of divisions, air wings, carrier task forces—is a responsibility of the services: Army, Navy, Air Force. And that is where the services have an opportunity to exercise creativity and initiative. The Army after the Civil War, driven to draconian measures by Congressional restrictions on manpower, all but abandoned larger formations—divisions, corps, armies. Operations during those grim, lean years were in the hands of some very capable, if understrength, regiments of infantry, artillery, and cavalry. It is a paradigm that we would do well to study; Tom Clancy's superb book is a good place to begin.

Donn Starry, General, U.S. Army, Retired,
41st Colonel of the Blackhorse, and
Honorary Colonel of the Regiment

Introduction

When did mobile warfare start? That's hard to say—but probably not long after somebody realized it was possible to use a horse to move things or people. And it was definitely going strong on the steppes of Central Asia by the third millennium BC. Recent excavations by Russian archaeologists of Bronze Age grave sites on the Kazakh steppes (dated around 2200 to 1800 BC) have unearthed the earliest known remains of chariots. These were invented as high-tech platforms from which warriors could shoot arrows or hurl javelins.

And yet it's quite possible that mobile warfare goes farther back than that. Bones from even earlier sites in the Ukraine suggest that the long love affair between humans and horses may have started more than six thousand years ago. Archaeologists debate the issue, but horses may have been ridden bareback long before they were harnessed to wheeled vehicles. What if the first use of the horse in battle was for reconnaissance? Sitting astride a horse you can see farther than you can while standing on your own two feet. And the horse has four legs, which has advantages, too. More fleet of foot than a man—though only for short distances, and only if properly treated—the horse can give his rider the ability to locate the enemy, approach him, count his numbers, perhaps harass him a little, and then escape unhurt to report to the chieftain. And so from time immemorial, these two missions have been the main missions of the cavalry: to locate the enemy, and to sting him.

Cavalry has rarely been a decisive arm by itself. For one thing, the size of the horse gave cavalry troopers lower combat density than the infantry. The breadth of a horse's chest and the space needed to avoid crushing a rider's legs against his neighbor's mount meant that two or three infantrymen occupied the same frontage as a single horse and rider. Two or three spears, swords, or bows in the hands of foot soldiers confronted each warrior on horseback. Less appreciated is a horse's unwillingness to plunge headlong into a barrier it cannot see through. Though a horse might not be the smartest living thing on earth, only men will knowingly hurl away their lives. Third, a horse is not a machine. To operate and perform properly, it needs food, water, and rest. Denied those things, it dies; and all the spare parts in an Army inventory can't fix that. And so it was a rule of the American West that on any long-distance trip of more than five days, an infantry company could outmarch a cavalry troop. A horse afforded a trooper a relatively high dash-speed, but only over fairly short distances. A man sitting on a horse also made an easy target, especially after the development of firearms. And yet, despite these drawbacks, the horse remained important in war for three millennia. More precisely, the horseman performed several crucial missions: find the enemy; prevent the enemy from finding you; collect information on the enemy before your main force collides with his; harass his flanks and communications; pursue him in defeat; screen your own forces when you are forced to withdraw.

Today the horse is used mainly for parades and ceremonies, but the missions it once performed remain as vital as ever. Though today's cavalry "companies" are called "troops," and the "battalions" are called "squadrons," the troopers (otherwise called "soldiers"—traditions do die hard, especially when John Ford made so many great movies about the glorious horse-soldiers) ride to battle not on Front Royal remounts, but mostly within sophisticated fighting vehicles.

Always the Army's proud arm, the socially prominent arm, the "pretty" arm—and for all those reasons despised by the infantry—the United States Cavalry[1] is not—and never was—just fashionable. It grows and changes. And so in the 1950s and '60s it mutated into a shock-arm. In those days, the 11th Armored Cavalry Regiment (ACR)[2] was tasked with covering the Fulda Gap, an historic invasion route into western Germany. The job of the 11th ACR was to slow down, break up, and generally obstruct the advance of an armored formation as large as the Soviet Third Shock Army (about twelve times its size). That job demanded a new kind of unit, different from one designed for reconnaissance. Consequently the armored cavalry regiment evolved into something like an unusually robust brigade, or even a mini-division—a superbly balanced combat formation, containing a little bit of everything the Army has, under the command of a full colonel. In due course, the ACR became a plum assignment, where successful stewardship was the passage to greater things. In fact, the top ranks of the U.S. Army are packed with men who have served in, and commanded, the three ACRs that operated during the Cold War.

This growth process, whose purpose was simply to give the unit designated to be the first target for the Red Army a modest chance at survival, ended up producing a military organization with unusual relevance for the world that is now emerging after the fall of Communism. Relatively small in size, the ACR is heavy on "teeth" and short on "tail"—a weighted fist with deceptive agility on the battlefield. It has global mobility, and the greatest concentration of firepower of any land combat force yet created. As we will see, the marriage of weapons and mobility, added to the coming revolution in battlefield-information technology, will transform the ACR yet again into a form that will make it the most important land component in the U.S. military's continuing mission of keeping the peace—and punishing those who violate it.

And that will continue to be the legacy of those who stir to the sound of "Boots and Saddles."

Tom Clancy
Peregrine Cliff, December 1993

ARMORED CAV

A Guided Tour of an Armored Cavalry Regiment

Armor 101: An Armored Warfare Science Primer

Ever since the first warrior padded himself with leather to ward off the blows of a rival's weapon, there has been an endless struggle between those who fashion armor to protect soldiers and those who build weapons to penetrate and destroy it. Later, when man began to forge metal into plates, he beat it into improved armor for his breast and head, to better ward off enemy spears and arrows. A well-armored warrior could close with his enemies, survive their attacks, and then destroy them with his own weapons. When sharp iron is flying around a battlefield, armor protection can be the difference between life and death, victory and defeat.

The knights of France fell to the archers of King Edward III of England at the Battle of Crécy (1346) because the armor a horse and rider could bear did not stop the arrows fired by longbows. The first British tanks were impervious to the machine-gun bullets fired by German soldiers during the Battle of Cambrai in 1917. The world of the armored warrior is an ever-shifting balance between armor and firepower. The way these two elements interact, as well as the skill of the operators, determines how well any armored fighting vehicle will do in combat. Let's take a quick look at the state of the art in the complex science of armored warfare.

Armor—The Hard Shell

Armor is the tank's reason for being, not mobility or a big gun, although both are desirable and will compete with armor in a tank's design. Armor is designed to keep the crew, and the weapons capable of inflicting punishment upon an enemy, safe.

Tanks were born in World War I out of the desperation of trench warfare. In September of 1914, after the Battle of the Marne, the defeated German Army fell back and dug into a system of trenches and defensive positions so strong that the Allies could not dislodge them. In the years that followed, machine guns and artillery inflicted hideous losses on both sides; the war bogged down into a stalemate. The addition of several hundreds of miles of barbed wire created a deadly zone between the opposing armies (called "no-man's-land") where infantry and the now-outdated horse cavalry (they just didn't know it yet) were mowed down by the tens of thousands. Allied commanders demanded more and more artillery to beat down the German troops and destroy their barbed wire and entrenchments. That didn't work. The only result was to turn a good portion of northern France and Belgium into terrain resembling a barren moonscape.

Something new was needed. Something that could crawl right up to the machine-gun nests and destroy them, without being destroyed first. The solution was a combination of steel plate (from naval ship armor), the internal combustion engine, caterpillar tracks (from early agricultural tractors), and machine guns or light cannon. Enter the tank. Because they were originally called "land ships," many terms in the tank vocabulary—turret, hull, hatch, deck, periscope—are naval metaphors, but not the name "tank" itself. That comes from a British cover story: They concealed their construction from the Germans by calling them storage tanks or boilers. During World War I the typical Allied tank had armor between 10mm (about .4") and 25mm (about 1") thick of hardened steel plate. Only a 12mm (about .5") thickness of armor was sufficient to stop German armor-piercing bullets at point-blank range. This was also sufficient to stop most artillery shell fragments, although a direct hit was usually fatal.

By the beginning of World War II, tank armor was between 30mm (about 1.2") and 70mm (about 2.75") thick; and the front section was sloped for improved penetration resistance. Unfortunately, with armor this thick it was no longer possible to provide all-around protection and allow the tank to move at a reasonable speed. Engineers therefore began designing tanks with a heavily armored front, while side and rear armor was only about half the thickness of the armor up front. During the war, tank designs and technology improved at a rapid pace; and by 1945, tank frontal armor ranged from 100mm (about 3.9") to 150mm (about 5.9") in thickness, although some German designs sported front armor between 200mm (about 7.9") and 240mm (about 9.4") thick (thicker than the armor on naval heavy cruisers).

Postwar tank designs followed these trends, with all then having frontal armor thicknesses in the 100mm (about 3.9") to 120mm (about 4.75") range. There were, however, significant improvements in the size and power of main guns. Bore size (the diameter of the shell fired by the gun) had now grown to between 100mm (about 3.9") and 115mm (about 4.5"), as compared to the 75mm (about 2.9") to 90mm (about 3.5") that was the average during the last years of World War II. Meanwhile, short-range anti-tank launchers like the American "bazooka" and German Panzerfaust gave infantry a tank-killing weapon.

By the early 1960s, frontal tank armor had started to increase in thickness again, mainly in response to the larger penetration capabilities of the new High-Explosive Anti-Tank (HEAT) projectiles that were appearing on tank-gun and recoilless-rifle rounds, as well as the first of the new Anti-Tank Guided Missiles (ATGMs). It is also during this time that the composition of armor began to change, with new designs no longer consisting simply of steel in ever increasing thicknesses. Instead, armor now combined steel with ceramic laminates to provide a tougher target for the new types of anti-tank weapons being deployed. This combination armor was first deployed on the Soviet T-64 main battle tank (MBT), which entered production in 1967. Unfortunately, because of the Vietnam War, the United States fell behind in armor development. So costly was the continuing conflict in Southeast Asia, that the U.S. Army missed an entire equipment modernization cycle. Frontline Army units continued to use tanks based on 1950s technology (the M48 and M60 series) well into the 1980s.

Modern Tank Armor

So what makes up the armor of a modern tank or fighting vehicle? Three ingredients determine just how effective a tank's armor-protection system, or package, will be. They are:

- The thickness of the armor package.
- The material composition of the armor package.
- The slope angle of the armor package outer face relative to an incoming weapon.

We'll examine how each of the three ingredients contributes to overall armor effectiveness and how they work together against various forms of attack. After that, we'll look at the exotic new generation of explosive reactive armor (ERA) and how that changes the picture.

Armor Thickness—Where armor is concerned, there is an old design axiom that thicker is better. Though the reason for this is intuitively obvious, I need to explain how that axiom is valid today, and how it has changed. All modern anti-tank weapons, with the exception of a few mines, use some sort of a penetrator to pierce the tank's armored hide and cause havoc inside the vehicle. The more material a penetrator has to work its way through, the lower the chance of a lethal penetration. But there is a practical limit to how much armor a tank can have and still move across terrain and carry a useful weapon. A large solid block of steel is fairly safe from penetration, but it just sits there and doesn't do anything.

The classic material for tank armor is a family of high-quality steel alloys, rolled to provide a uniform thickness as well as the best combination of strength and hardness. Because the material is of uniform hardness throughout, this type of armor is called Rolled Homogenous Armor or RHA, and it is the standard by which all armor types are judged. Though many factors, other than just thickness, go into estimating the RHA equivalent (measured in millimeters of RHA armor plate) for a given type of armor, this simple numerical rating allows all types of armor to be evaluated comparatively. For example, the 1943-vintage M4 Sherman tank that my wife Wanda gave me for Christmas a few years back has an RHA equivalent thickness of 100mm (3.9"). By comparison, the first M1 Abrams tanks delivered in the early 1980s had an RHA equivalent value of almost 450mm (about 17.7") against a kinetic-energy penetrator (solid-shot). And the current version of the Abrams, the

M1A2, has an RHA equivalent value of almost 800mm (about 31.5") against a kinetic-energy penetrator, and an amazing 1,300mm (almost 51.2") against HEAT-type weapons!

Materials—More goes into an armor-protection package than a simple thickness of steel or other material. Indeed, the composition of the package has a great bearing on just how much protection it provides. Modern armor designs are complex combinations of materials (steel, ceramics, exotic metal alloys, and even plastics). For example, the Chobham armor used in the early M1/M1A1 tanks is far more effective against HEAT (chemical-energy/explosive) rounds than against long-rod (solid-shot) penetrators. Thus the M1A1 Heavy Armor (HA) variant, with a layer of depleted-uranium armor, was designed primarily to defeat long-rod penetrators.

By now all this talk of HEAT rounds and long-rod penetrators may have you wondering what I am referring to. So perhaps some explanation is necessary before we go on. HEAT—or shaped-charge—rounds had their origin in the Second World War when weapon makers borrowed an old miners' trick to "shape" or focus the energy from an explosion into a small area so it could penetrate armored plate. By the end of WWII, such munitions became a serious threat to the tank. And in the early 1960s, when a shaped-charge round was mated to a rocket motor and a guidance system, a really practical, lightweight tank killer had come into being—the anti-tank guided missile (ATGM). Now small vehicles and infantry had the ability to attack and defeat a tank's frontal armor. Such an attack would have been suicidal in the past. The ATGM proved itself in the 1973 Arab-Israeli War, when several Israeli armored brigades, without infantry support, were badly chopped up when they attacked Egyptian infantry positions (infantry protects tanks by spotting enemy missile teams and suppressing them). During the mid-1970s, military experts debated whether or not ATGMs had rendered the main battle tank obsolete.

But some officers in the U.S. Army chose to look upon the data of the '73 war as more than an epitaph for the tank. A closer look at the operational data from the war showed Western tank designers that existing armor types, usually a thin "face-hardened" surface layer on a thick plate of RHA, did not provide sufficient protection against the current generation of HEAT warheads, let alone the next generation due to appear in the early 1980s. As we mentioned earlier, the Soviet Union was fielding tanks in the 1960s with

A diagram of a HEAT round impacting a piece of sloped armor. The round on the left is shown just prior to impact; the one on the right just after detonation of the shaped charge.

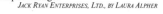
JACK RYAN ENTERPRISES, LTD., BY LAURA ALPHER

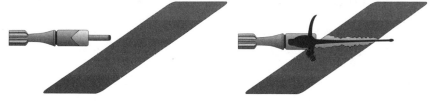

combination (metal and ceramic) armor, long before their Western opponents. But in the early 1970s, the British Army research facility at Chobham, England, developed a revolutionary type of armor with a honeycombed ceramic composite layer sandwiched between steel plates. This new material combination was code-named Burlington, but it is better known by its trade name—Chobham armor.

Because of their mechanical properties, mixed layers of ceramic and metal composites offer excellent protection against HEAT rounds. Ceramics are amorphous, that is to say that they do not have a crystalline structure like metals, but are more "fluid-like," with a fairly random molecular structure. So when a HEAT round strikes combination armor, the explosive jet quickly pushes through the outer steel layer and tries to burrow through the ceramic. However, unlike metals, which break up along structural boundaries between individual crystals (which stay separated), ceramics tend to flow around the jet and break it up into many smaller "jetlets," which are quickly dissipated. The disadvantage of combination armor is that it is very bulky, since the ceramic and composite layers have to be of sufficient depth to break up the jet. Consequently, although ceramics are lighter than metals, the overall weight (or mass) of a combination-armor package tends to be about the same as RHA for a given level of protection. Examples of its use can be found in the turret designs on the American M1 Abrams and the British Challenger tanks that were employed so successfully during the 1991 Persian Gulf War. In fact, the newest Chobham-type armor designs tend to be about two to two-and-a-half times more effective against HEAT-type weapons than an equivalent weight of solid RHA.

By the early 1980s, as tanks with combination armor entered service, HEAT rounds became less of a threat to tank survival. This made the kinetic-energy penetrator, once again, the main tank killer. Kinetic-energy penetrators rely on impact to pierce armor. While HEAT rounds derive most of their penetration capability from the explosive jet's velocity, kinetic-energy penetrators use both mass and velocity to do the job. Modern armor-piercing fin-stabilized discarding-sabot (APFSDS) rounds are very dense, long, slender darts (hence the name long-rod penetrators) that burrow into a tank's armor on impact. If it has sufficient kinetic energy, the dart goes right through the tank's armor and raises hell inside. These long-rod penetrators, unlike shaped-charge jets, are solid and are not broken up by the amorphous structure of the ceramics in combination armor. However, ceramics possess another property that has an effect on kinetic-energy rounds. This property is hardness: the ability of a material to resist scratching or penetration. The higher the hardness value, the greater the kinetic energy required to penetrate the material.

Now, as far as kinetic-energy protection is concerned, RHA is pretty soft stuff (relatively speaking), and can easily be pushed aside by a high-velocity long-rod penetrator. Though steel can be hardened by special treatments (such as nitriding or carburizing), you can't make an entire hull out of very hard steel. It is exceedingly difficult to fabricate and weld large thick pieces of high-

An APFSDS round impacting and defeating an armor plate. Note the spall fragments which are thrown inward by the penetration of the "dart." *JACK RYAN ENTERPRISES, LTD., BY LAURA ALPHER*

hardness steel into armored structures. Another problem is that metals with high hardness are brittle and tend to shatter like glass under high-energy impact. So combining layers of hard and soft steel is advantageous. If the hardened face steel is tough enough, the attacking projectile may be deflected outright, break up on impact, or have its nose squashed (i.e., no longer a nice sharp point). Because of this "blunting" effect, the projectile has to expend more energy to penetrate the underlying softer steel. And the softer steel can absorb energy because it readily deforms, or gives, under the load. In this way it tends to dissipate the energy of the round.

Hardened steel, however, is still relatively soft compared to ceramics. For example, silicon carbide (a ceramic used to make drill bits) is about three to four times as hard as RHA. Thus, combination armor with hard ceramic/composite blocks supported by RHA protects against kinetic-energy attacks about as well as a combination of hard and soft steels (one needs to remember that these armors were specifically designed to defeat the large HEAT warheads in ATGMs). Another advantage (sort of a backhanded advantage actually) that combination armor has over RHA is that it is usually thicker, so there is more material that a long-rod penetrator has to go through before it gets to the inside of the tank. And the addition of a layer of depleted uranium (DU) in the HA variant of the M1 Abrams tank makes it even more resistant to kinetic-energy rounds. While precisely how this is done remains one of the *really* black secrets within the U.S. Army, it doubles the effectiveness of the M1's armor against long-rods. The exact composition of the M1's armor is so secret that the armor package stuffing room is the only place that you are not allowed to see in the M1 factory at Lima, Ohio. In fact, one worker on the factory floor described it as the "kryptonite" room.

Armor Slope—A final factor in the effectiveness of an armor package is the slope of the armored face. The degree of slope of an armored surface has two major effects. First, if the slope angle is 60° or greater, there is a good chance that a projectile will ricochet off the armor face, causing little or no damage. Second, the slope angle determines the amount of armor that a long-rod penetrator or HEAT jet will actually have to push through before it reaches the tank's interior. The basic rule of thumb is that the greater the slope, the greater the protection; for that increases the effective thickness of the armor. Thus, the average frontal armor slope has increased steadily throughout the history of the tank. In World War I the slope was 0° on many vehicles.

By the end of World War II, frontal armor slopes were between 45° and 60°. Modern main battle tanks generally have front armor slopes around 70°, although the M1 series of tanks have a front armor slope of about 80°. Thus, if a tank has a frontal armor plate 200mm thick, with a slope of 70°, then the actual thickness a weapon would have to defeat would be 584mm. This is a lot of armor!

So how do all of these things add up when it comes to a real-world tank design? Consider the following. The Russian T-72 tank (according to reports published in defense journals) has frontal armor made up of a layered combination of steel, ceramics, and composites. Its thickness is about 200mm, and the slope is 68°. A rough calculation of the RHA equivalents for chemical and kinetic-energy attacks would give an RHA equivalent of 720mm (about 28.3") against a HEAT round, and 454mm (about 17.9") against a kinetic-energy penetrator. The danger with this type of calculation is that each armor type can't be easily boiled down to a simple number. And there is little published in the open literature about how different types of armor arrays react to each type of weapon. What this crude estimate of the T-72 does illustrate, though, is that it probably has sufficient protection to withstand an attack from almost all infantry anti-tank weapons with HEAT rounds (400-600mm penetration capability) such as the Russian RPG-7. In addition, penetration isn't an end in itself. There must be sufficient residual energy in the HEAT jet to disable or kill a tank and its occupants. If an ATGM has just enough energy to penetrate the armor, the tank will still be able to fight. Thus, there has to be some residual energy left to blow fragments and spalls (chunks literally peeled off the armor) into the interior in order to kill the tank or the crew.

Reactive Armor—The newest fashion in armor technology is explosive reactive armor (ERA). ERA was developed by Israel (under the trade name of Blazer) and deployed on Israeli Merkava, as well as the U.S.-made M60 and M48 Patton tanks, during the 1982 Lebanon invasion. During that operation, the Israeli Army lost several tanks fitted with ERA; and it is rumored that the Arabs gave samples to the Soviets. In any event, by about 1985, Soviet T-64, T-72, and T-80 tanks started showing up with heavy applications of ERA on the front, sides, and turret. By adding the ERA to their tanks (already fitted with HEAT-resistant combination armor), the Soviets had sufficient protection against virtually all current ATGMs.

ERA looks like a set of child's toy blocks, and is usually fitted to the exposed forward surfaces of an armored vehicle. It works like this: When a shaped-charge jet hits an ERA block, the impact detonates an explosive charge sandwiched between two steel plates. This explosion pushes the outer plate into the HEAT jet, which now has to cut through the plate to get to the main armor. The inner plate is driven against the hull and rebounds back into the path of the now-disrupted jet. The energy of the shaped-charge jet is absorbed by the two moving plates, as the jet must continually cut through fresh material. The remaining highly disrupted jet lacks sufficient punch to penetrate the main armor.

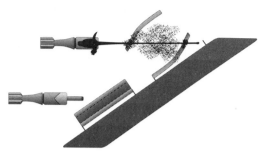

Reactive armor in action. In the lower view, a HEAT round is approaching a reactive-armor box (a sandwich of explosive between metal plates). When the round strikes the box (see the upper view), the explosive in the box detonates at the same time as the HEAT warhead, disrupting the flow of the plasma jet.

ERA does have some severe limitations, however. The effectiveness of ERA against shaped-charge jets is very dependent on slope angle. When you look at a Soviet/Russian T-72 fitted with ERA blocks, you see these at roughly the same angle as the main hull. At an angle of 68°, the ERA disrupts approximately 75% of the shaped-charge jet, but ERA at an angle of 0° only disrupts about 10–15% of a HEAT jet. It didn't take long to figure out that top attack (a warhead firing down into the tank) was one way to defeat this new armor variation. Even if you completely cover the top of your turret with ERA, a top-attack warhead will strike it at an angle close to 0°, so you gain little extra protection for the cost and weight. The Swedish Bill and the U.S. TOW-2B are examples of this technique in use. Another variation is the "dual" warhead, which uses a small explosive charge to detonate the reactive-armor block. This is followed by a main warhead to defeat the primary armor of the tank. Such a dual warhead is used in the U.S. Army's latest model of Hellfire anti-armor missiles. The Russians have even designed a triple-warhead, 125mm HEAT round, designated 3BK27, which is reportedly able to defeat modern Western armor packages.

Earlier I mentioned that a long-rod penetrator would also detonate the ERA block. However, unlike the HEAT round explosive jet, which has very little mass, the long-rod is just too massive for the thin steel plates of the ERA block. As a result, penetration is only marginally degraded by going through a layer of ERA. There is on the drawing boards an ERA block with thicker plates, designed to snap the long-rod penetrator in two through a shearing action. While this "thick-wall" ERA gives improved protection against long-rods, the protection it gives against a HEAT round is reduced. There is no such thing as a free lunch!

Using ERA poses two more problems. First, it is one-shot protection. Once a spot on an ERA-protected vehicle is hit, the area that has been struck is no longer shielded (until you install a new ERA block). Second, dismounted infantry cannot escort tanks fitted with ERA, because the exploding ERA blocks throw out lots of fragments, shredding any nearby troops!

In other words, modern main battle tanks have *not* become science-fiction ogres—irresistible killing machines. For one thing, it must be remembered that today's tanks and armored fighting vehicles only have massive protection on the front of the vehicle. The sides, rear, top, and bottom are not only relatively thin-skinned, but there is no way to angle the armor in these locations effectively. So, while a T-72 with reactive armor has an RHA equivalent of over 1,000mm (about 39.4") against a HEAT round on its front, it has at best between 100mm (about 3.9") to 135mm (about 5.3") of RHA equivalent armor on the sides, and even less in the rear. Thus a tank is anything but invulnerable. And only good tactical employment will keep its strong side aimed toward the weapons that are trying to destroy it.

Consider the following: The Hellfire antitank missile, originally fielded in 1986, had about 1,050mm of RHA penetration capability. It probably would not penetrate a Soviet T-72 tank's front plate augmented with reactive armor. But if fired into the T-72's side, a German World War II Panzerfaust 100 (fielded in 1944, with about 200mm RHA penetration capability) would cut right through and probably knock the tank out. Hellfires fired at T-72s from the side during Desert Storm usually blew the tank's turret into the air, and blew out the sides of the armored box hull. Thus, the ogre has Achilles' heels.

And this brings up a point about Soviet tank design. Compared to tank designs in the West, Russian tanks have small front profiles. Russian tanks are designed to present the smallest possible target, reducing the probability of being hit. When they are trying to sell their tanks to other countries, the Russians promote this as a significant advantage. In comparison to the German Leopard II, for instance, the T-72 is roughly half a meter (about 20") shorter in height and three-quarters of a meter (about 30") shorter in length. Additionally, compared with the massive box-like turrets of the German Leopard II and the American M1 Abrams, Russian turrets are also very small. Unfortunately, users of Soviet tanks like the T-72 have found that the relatively small reduction in target area means a severely reduced internal volume. In comparison to Western tanks, Russian tanks have about half the usable internal space. Should a long-rod penetrator or a shaped-charge explosive jet penetrate the hull or turret, there is a higher probability that fragments and spall will hit something critical. There is simply less space to rattle around in. And with such small turrets, the Russians have no room to store ammunition in blast-resistant bustles with blowout panels (such as the American Abrams tanks employ). Instead they have to store ammunition down in the hull, risking catastrophic explosion if the armor is penetrated. So if a Russian used-tank salesman offers you a deal on some low-mileage T-64s or T-72s, just say *Nyet!*

Anti-Tank Weapons—The Dragon Killers

Over the last three-quarters of a century, a variety of different weapons has evolved to destroy the armored monsters that roam the battlefield. The first anti-tank weapons were large-caliber rifles (similar to those used for hunting elephants and rhinos) firing heavy slugs. These crude weapons were soon replaced by more effective methods of penetrating armor. Historically, the most dangerous enemy of a tank is another tank. The projectiles thrown by tank main guns have become the most lethal anti-tank weapon. To understand why, let's look at a few of these dragon killers.

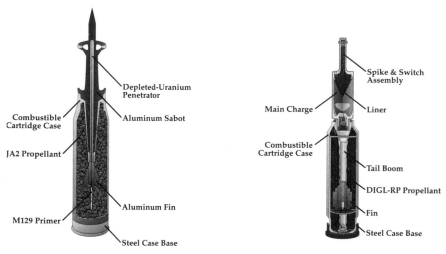

LEFT: A cutaway of 120mm M829 kinetic-energy penetrating round of the type used on the M1 Abrams main battle tank. A special version of this round, the MY29A2, became known as the "Silver Bullet" during the 1991 Persian Gulf War.   JACK RYAN ENTERPRISES, LTD., BY LAURA ALPHER

RIGHT: The 120mm M830 HEAT round used by the M1 Abrams main battle tank.
JACK RYAN ENTERPRISES, LTD., BY LAURA ALPHER

Long-Rod Penetrators—Long-rod penetrators are officially known as High-Velocity, Armor-Piercing, Fin-Stabilized, Discarding-Sabot (HVAPFSDS) projectiles. These are sub-caliber (smaller than the diameter of the firing gun) projectiles designed to penetrate a tank's armor through brute force. To put it another way, this weapon is the modern equivalent of King Edward's longbow archers at the Battle of Crécy. Modern long-rod penetrators are made of solid tungsten or depleted-uranium alloys. Both are extremely dense and very hard. Some earlier long-rods were made of stainless steel with tungsten cores, but these tended to shatter on impact with the modern armor packages used on today's tanks. Depleted-uranium (DU) alloys have slightly better penetration performance than tungsten alloys, but DU is somewhat radioactive and the dust (UO_2) is *very* toxic. Since tank ammunition is stowed in a separate armored compartment, the tank's crew is effectively shielded from radiation and chemical hazards before the DU round is fired, but this protection is lost when the round impacts on a target. Consequently, battlefield contamination has raised some environmental concerns. In addition, DU is difficult to manufacture and work with safely. So why bother with DU at all?

Well, first of all, DU penetrators perform a bit better than tungsten, and you never know when that little bit of extra performance will save your tank (and maybe your behind). Second, as a DU long-rod penetrates armor, the nose flakes into small particles. These particles become incandescent from the friction generated. When these particles are finally blasted into the tank's interior, they ignite and burn violently, generating heat and pressure that is deadly to the crew and the tank (with its stored fuel and ammunition). This "pyrophoric" effect gives DU an advantage over tungsten in lethality.

Present U.S. APFSDS rounds, like the M829A1, are made of DU alloy and have a length-to-diameter ratio of around fifteen or twenty to one. That is, the round length is fifteen or twenty times the round's diameter. But experimental APFSDS rounds, such as the M829A2 Desert Storm "Silver Bullet," are about thirty to one. Decreasing the long-rod's diameter improves its penetration capability. In short, the smaller-diameter penetrator forces a smaller area of armor to absorb the rod's kinetic energy. Reduced diameter also means lower drag, so that the thinner round impacts the target at a higher velocity, with a reduced time of flight. Clearly this is bad news for a tank! On the downside, the smaller-diameter penetrator is more likely to break up as impact stresses are applied to the smaller cross-sectional area.

The velocity of an APFSDS round as it leaves the muzzle is tremendous. For the M256 120mm smoothbore gun on the M1A1 and M1A2 variants, the muzzle velocity is approximately 1,650 meters per second (about 5,413 feet per second), or about Mach 4. Because of this high speed and the projectile's length, long-rod penetrators are fin-stabilized to prevent wobbling in flight. In essence, the long-rod penetrator is a 1.5-to-2-foot-long (.46-to-.61-meter-long), 10-lb. (4.54-kg.) metal dart. The need to accelerate the round to these speeds requires the use of a sabot, which holds the round in the middle of the gun barrel as it travels towards the breach. As the round exits the gun tube, aerodynamic drag tears the sabot from the penetrator, and the round flies on towards its target. This use of sabots—and the higher operating pressure and velocity of their rounds—is one of the reasons for the popularity of smoothbore tank cannons. Older rifled guns have spiral grooves in the barrel that impart a stabilizing spin to the round. But rifled guns cannot sustain the very high pressures and muzzle velocities of sabot rounds because the grooves would rapidly erode, making the tube useless after a few rounds.

Because of the slope and composition of modern armor, the APFSDS round must fly perfectly straight to penetrate even moderate armor. Any deviation from a stable trajectory and the APFSDS round could lose upwards of 80% of its penetrating power, or even snap in two under the enormous stress of impact. But if the long-rod penetrator hits a tank's armor squarely, the highly localized pressure will deform and push the armor material out of the projectile's path. In essence, the armor flows out of the way of the projectile as it travels and forms a penetration cavity. Once through the armor, the remnants of the APFSDS round and some armor fragments (spall) are injected into the tank's interior, with serious consequences for the occupants: They are literally sitting ducks in a big steel barrel. Many APFSDS rounds were fired during Desert Storm, with stunning results. In one action, a British Army colonel hit and destroyed an Iraqi tank with an APFSDS round at over 5,400 meters (about 3.35 miles). That is really long-range shooting!

HEAT/Shaped-Charge Rounds—To kill an armored target, you must defeat the armor by blowing a hole in it. Current weapons technology provides two ways to do this. You can use the kinetic energy of a long-rod traveling at very high speed, typically a mile/1.6 kilometers per second. Or you can use the chemical energy released by an explosive "shaped charge," which can be traveling quite slowly indeed.

HEAT rounds use a high-explosive shaped charge packed around a cone-shaped metallic liner. When detonated, the explosive causes the metal liner to collapse rapidly inward. The metal liner (usually composed of copper or aluminum) is heated and compressed by the energy of the explosion to form a jet with velocities as high as 8,000 to 9,000 meters per second (about 29,500 feet per second), or about Mach 25! The metal liner, however, is not a molten stream. It is still very solid, yet behaves like a fluid because of the tremendous pressures involved. Modern HEAT rounds have a long probe or a hollow nose cone containing the fusing mechanism. This provides a standoff distance between the warhead and the target's surface. This standoff allows the jet to reach its optimum form, which maximizes its penetration capability.

When a shaped-charge jet hits a tank's armor, the highly localized pressure deforms and pushes the armor material out of the jet's path, thus forming a penetration cavity. If you think this sounds similar to what a long-rod penetrator does, you are correct. In both cases the weapons use high local pressure to exceed the armor's mechanical strength. As the armor material is moved out of the way, the jet or the projectile occupies the space created and continues to burrow inward. In both cases the armor behaves like a fluid and flows around the intruding object. Shaped-charge munitions can penetrate approximately four to seven times the diameter of the warhead in millimeters of RHA, depending on the liner material and the type of target armor. When combination armor is involved, this rough rule doesn't apply, because this type of armor was specifically designed to defeat HEAT rounds.

Explosively Formed Projectiles/Top Attack—Explosively formed projectiles (EFPs)—or flying plates, as they are often called—are kissing cousins to the shaped-charge warhead. Like the shaped charge, a high explosive is used to deform a metallic liner into a projectile. But this is where the similarity ends. Shaped-charge warheads use a conical-shaped liner, whereas the EFP uses a shallow hemispherical dish. When the explosive detonates, it deforms the dish into a solid slug or penetrator, rather than an elongated jet. The slug is aerodynamically stable (like a long-rod) and can reach speeds up to about 2,000 meters per second (about 6,560 feet per second), or about Mach 5 (fairly slow compared to the 8,000 or 9,000 meters per second reached by most HEAT-round jets). Consequently, EFPs penetrate less armor than shaped charges: A rough rule of thumb for EFP penetration is millimeters of RHA equal to the warhead's diameter. (Recall that shaped-charge warheads can penetrate about four to seven times the warhead diameter of armor.) EFPs are used mainly in top-attack weapons or mines, where massive armor-penetration capability isn't required. Examples of such weapons are the Swedish Bill and the American TOW-2B.

Because they circumvent the tank's thickest armor, top-attack munitions have gained in popularity in recent times. In the case of an EFP slug against a T-72, the slug only has to penetrate about 200mm of top armor, not 454mm of frontal armor. Even tank main-gun rounds are being designed to take advantage of this indirect technique. Watch this technology, for it is the future.

There and Back Again: An Interview with General Fred Franks

Over twenty years ago, at the close of the Vietnam War, it was hard to find an organization held in less respect than the U.S. Army. An American force of over 500,000, most of them U.S. Army personnel, equipped with U.S. Army equipment, and trained to U.S. Army doctrine, had just done something unknown in the history of American arms. They had won every battle and engagement, but left Southeast Asia with their Air Force and Navy comrades without victory. The political reasons for the defeat in Vietnam were complex and controversial, but people still found it easy to blame the leadership, personnel, and equipment of "the big green machine" (rather than the politicians and bureaucrats who truly deserve the blame) for America's first lost war.

Yet, beneath this defeat and waste lay the first roots of the reborn Army that would crush Iraq in four days and set new standards for the art of warfare. With twenty years of hindsight, it is easy to see just how deeply those roots extend. Certainly to General Creighton Abrams, the Army Chief of Staff after Vietnam, who as a young officer in World War II led the relief column that first reached the besieged town of Bastogne during the Battle of the Bulge. In his mind was a dream of a new professional army.

There also was the professional corps of young officers that had served in Vietnam during the war. Those who chose to stay in the Army and "stick it out" surely made a promise to themselves: "If we ever reach the rank of general, things are going to be different!" Men like Colin Powell, who had overcome the racism of an Army struggling to master the hard lessons of integration. Armor officers like Butch Funk and Infantry officers like Pete Taylor, who came home with reputations as tough as their weapons. And guys like Barry McCaffrey and Fred Franks, who left bones and limbs on the killing fields of Southeast Asia.

Profoundly affected by the sacrifice, the loss, and the waste they had experienced, a generation of officers set out to rebuild the Army so that such a tragedy would never happen again. They did this as an act of honoring and keeping faith with those who were lost and maimed, to show that those soldiers had not suffered in vain. Thanks to their integrity, intelligence, and skill, America has an army which other nations not only envy and respect, but fear. Military analysts for decades pointed to the Israeli Defense Force as the model of how an army should work. Now, even the vaunted Israeli force which held the Golan (as I described in *The Sum of All Fears*) and invaded Egypt in 1973 seems at times to be considered a second-class force compared to the units that America can field.

General Fred Franks, Commander of the U.S. Army, Training and Doctrine Command (TRADOC), Ft. Monroe, Virginia. *OFFICIAL U.S. ARMY PHOTO*

To get a better idea just what the U.S. Army has gone through to reach this point, let's have a chat with General Frederick Franks. General Franks is currently the commanding officer of the U.S. Army's Training and Doctrine Command (TRADOC), headquartered at Fort Monroe, Virginia, beside the old fortress that overlooked the first battle between the ironclad vessels USS *Monitor* and CSS *Virginia* (better known as the *Merrimac*). Commanded by a four-star general, the Army's highest rank, TRADOC is responsible for the organization, tactics, and training of the U.S. Army. It is here that the Army's way of fighting is defined, documented, and disseminated to roughly one million U.S. Army, U.S. Army Reserve, and National Guard troops that together make up what is known as the "Total Force."

Frederick Melvin Franks, Jr., was born on November 1st, 1936, in West Lawn, Pennsylvania. After growing up in Pennsylvania and attending one year at Lehigh University, he entered the U.S. Military Academy at West Point, New York, on July 5th, 1955. He graduated on the 3rd of June, 1959, and was commissioned a second lieutenant of Armor. Let's let General Franks tell the story from there.

Tom Clancy: What was the first unit that you were assigned to?

General Franks: I had wanted to select the Armor branch and go into an armored cavalry unit—one that was fully engaged in the Cold War. I ended up going to Europe and serving in the 11th Armored Cavalry Regiment at Regensburg, Germany, beginning in March of 1960. I was assigned to the 3rd Squadron of the 11th.

Tom Clancy: What was it like when you arrived?

General Franks: It was . . . great duty. The regiment had an important mission up on the Czechoslovakian border with considerable early responsibility. There was also a lot of opportunity for a young lieutenant to grow professionally. We had several WWII and Korean War veterans who provided valuable experience and insights for young officers, NCOs, and so many other young men like myself.

Tom Clancy: What was the Army like in those days? Lifestyle and so on?

General Franks: The unit I joined had just recently replaced the 6th Armored Cavalry Regiment in what we then called a "Gyroscope" rotation. Though I did not think so at the time, as I look back on it, we [the Army] were in a transition, and we were on the way up. My senior officers and NCOs had mostly fought in World War II and/or the Korean War, and the draft was also running in those days. Though there was a lot of confidence in the abilities of the unit, nevertheless, because of the mix of older, more experienced soldiers, younger officers and NCOs, and the pool of troops brought in by the draft, we had a different chemistry and dynamic from anything you might find in the Army today. All the same, the 11th Cavalry was a spirited outfit, and we were focused on a "real-world" mission, tense with all the various crises [Berlin and Cuba in particular] at the time. We were in the field, away from home, and on the border a lot, looking "eyeball-to-eyeball" with our probable opponents. It was a time of very focused and intense effort.

Tom Clancy: Somewhat on "the brink"?

General Franks: We had the Berlin Crisis in '61 and the Cuban Missile Crisis in '62. I personally felt very fortunate to have a lot of responsibility very early in my career, and the intense operational environment made you grow professionally very quickly. I learned from some great NCOs and some dedicated soldiers.

Tom Clancy: What was your equipment like in those days?

General Franks: We had a lot of World War II equipment mixed with some newer systems. Most of our personal weapons were of World War II vintage. I recall also that we had a big armored personnel carrier [APC] called the M59 that had two 2-1/2-ton truck engines mounted on either side of the vehicle. It was a maintenance nightmare, and the engines never ran correctly. On a more positive note, we had the M48A2 Patton tank, which was the top-of-the-line model, with improved fuel injection on the engine and a 90mm cannon—though we were more than a little concerned that the latest Russian tanks (the T-54/55) had a 100mm cannon as compared to the 90mm cannon on our M48s. Our tank retrievers were mounted on old World War II-vintage M4 Sherman tank chassis. We also had a mixture of weapons like Browning Automatic Rifles (BARs), M2 .50-caliber machine guns, and *really* old M1919A4 machine guns mounted on the scout jeeps. Certainly in 1960, we did not have the high-tech, modern-equipment force that defeated the Iraqi Army in 1991.

The 1960s brought on significant changes for the U.S. Army, as the focus of American foreign policy switched from Europe and Cuba to the Far East and Vietnam. These started the armed forces of the United States in general, and the U.S. Army specifically, on a long slide that would not stop until the doldrums of the 1970s. The war itself brought on a time of great personal pain for Fred Franks, as he lost most of his left leg below the knee to enemy fire during the invasion of Cambodia in 1970. As it turned out, his climb back to becoming an effective soldier paralleled the rebirth of the Army as it struggled to come back from the nightmare of Vietnam.

Tom Clancy: By the mid-1960s, what changes were you seeing in the Army as we shifted to operations in Southeast Asia?

General Franks: In those days we had an Army that was very focused on basic fundamentals like gunnery skills, maneuver skills, and maintenance skills. We worked hard at it. And we had an Army that was confident in itself and that was looking to the future—the experimental placing of a division in the air-assault role [the 1st Air Cavalry Division] being a prime example.[1] It was most of all a deployable field Army, which had its "head in the game." Although I was not there then, as I look back, it seems to me that the Army we went into Vietnam with [in 1965 and 1966] was a highly proficient tactical war-fighting force. And the early tactical battles [in Vietnam], where our soldiers fought and won some tough engagements, bear that out. In many of the first battles in the wars our nation has fought, our Army hasn't always been ready. But in Vietnam, our force was tactically ready for those early battles. The units that we sent over early—units like the 1st Air Cavalry Division, the 11th Armored Cavalry Regiment, et cetera—were excellent in spirit and in the tactical proficiency they brought to the battlefield. General Hal Moore's recent account of battle in the Ia Drang Valley, and the toughness and heroism of our soldiers there, shows this dramatically.

Tom Clancy: What exactly went wrong for the Army in Vietnam?

General Franks: That's hard to sum up. . . . I served there in '69 and '70 as a major with the 11th Cavalry, and later went on into Cambodia with them, so my view was very narrow at the time. . . . A variety of things went wrong actually. We were putting together tactical victories in battles and engagements, but were lacking in some sort of coherent plan and pattern to achieve a strategic objective in the overall war effort. We lacked what we now call "operational art." I recall that there was a lot of communications back and forth between the military and civilian leadership as to what the strategic objectives should be and what the operational scheme of maneuver should be, and in general, just how things should be done. And unfortunately, this just went on and on through the '60s and on into the '70s.

Tom Clancy: What was happening to the Army at an institutional level?

General Franks: In my view, the loss of experience from officers and NCOs being killed, wounded, or rotated out resulted in the gradual loss of unit leadership and cohesion over a couple of years. As far as replacements went, we as a country did not call up the Reserves, and this was a major mistake in my opinion. Therefore, the active Army had to find replacement leaders wherever possible, and did so the best it could with the leadership pool that it had. But when the battlefield losses began to mount, the stress on the personnel system began to fracture the coherence of some of the units in Vietnam. The civil disagreements in our own country over the aims of the war, combined with the combat losses and the one-year DEROS [Date of Expected Return from Overseas Service] rotation schedule, really caused the Army significant difficulties in the early to mid-'70s. As a result of all these things, the Army as an institution suffered. This became a major problem as we began the transition to an all-volunteer force in '73, and had to start recruiting new people to serve.

Tom Clancy: What happened to you during your tour in Southeast Asia?

General Franks: First, I was privileged to serve in a great outfit with real heroes—2nd Squadron, 11th Cavalry. The end of my tour, as you know, was very painful for me personally, as I lost a portion of my [left] leg as a result of being wounded in action at Snoul, Cambodia, on May 5, 1970. And like so many others, I also lost some great friends. I then spent some time in an amputee ward in an Army hospital in Valley Forge, Pennsylvania.

I remember what impressed me most about my fellow patients was that they were magnificent young Americans who went and did what their country asked them to do on the battlefield. Now you can debate the merits of what they were asked to do; but one consequence of this debate was that for a long time afterwards, the public was not able to separate the war from the warriors. Unfortunately, the soldiers got caught up in all that. Back in the amputee ward at Valley Forge, my fellow soldiers got so tired of going back home and explaining what had happened to them, especially when they kept hearing what a waste all their experiences had been, that eventually they began to tell people they had been injured in things like factory explosions or auto accidents. . . . Yet while I was there, I do not recall a single elected official or an officer senior over the rank of colonel ever coming by to say "thanks" to these great young Americans.

Tom Clancy: After your experiences in Southeast Asia, did you make yourself any promises about your future service in the Army?

General Franks: Absolutely! I have not earned the right to speak for everyone, but I can speak for my own experiences. There's a hot blue flame that burns in my stomach about that time and what happened. I remember the soldiers I was privileged to serve with on the battlefield. In silence, you can feel the names on the Wall [the Vietnam Veterans Memorial in Washington, D.C.]. I also remember my fellow soldiers on the amputee ward at Valley Forge, and the sacrifice and the broken trust. As a result, I made a pledge to them, to myself, and a pledge to the Army, that if I had the opportunity to remain on active duty, I would do whatever I could in my circles of influence and responsibility, wherever they happened to be, to *personally* see to it that such things as broken trust would *never* happen again. Never.

Tom Clancy: What was your position in the Army when you got out of the hospital at Valley Forge, and where did you go?

General Franks: I was a major then, and I asked the Army to allow me to stay on active duty in the Armor branch. All I ever wanted to be was a soldier, and to be able to compete for positions in the Army. If my physical state should hold me back, so be it. But don't automatically assume that I cannot do the job just because most of what was below my left knee is gone. It's not what you don't have but what you do have that counts! Some senior leaders around the Army then worked hard to see to it that those of us with otherwise disqualifying physical profiles could remain on active duty. As a consequence of that, the Army allowed me to stay in the Armor branch, and I headed back out into the world. I remain grateful to the Army as an institution for that opportunity.

For my next assignment, I went to the Armed Forces Staff College [AFSC] in Norfolk, Virginia. But before that, my wife, daughter, and I went to Disney World in Florida. I will always have a soft spot for that place; it was the first time I could really walk after several rather severe operations. While at AFSC, I kept working out to stay in shape. I played softball and walked a lot. Rode a bike all around the AFSC. I also remember joking with my contemporaries about "having a leg up." After AFSC, I was assigned to the Department of the Army Staff in Washington, D.C. I worked there a year, and then worked for two years as the military assistant to the Under Secretary of the Army. I then got the opportunity to command the 1st Squadron of the 3rd Armored Cavalry Regiment at Fort Bliss, Texas, in June of 1975.

When I went to command the cavalry squadron, I'd been out of the Valley Forge amputee ward for a little over three years. As you might imagine, when I took command, I was a little tense. But I have to say that it was my fellow troopers and NCOs who really made me whole again as a soldier. Never once that I knew of was there ever any concern expressed about my being an amputee. They took me as I was, based upon what I could do, and that became a great source of strength for me personally. I'll never forget that family I was a part of for a year and a half. That experience completed my own personal climb back to being a whole soldier again.

You know, you derive inspiration from a number of places during a time of recovery. My family. My wife, Denise, and daughter, Margie. Both were inspirations and still are. They know the difference between compassion and pity. My mother and dad, the skillful doctors and medical staff of the Army Medical Service—including the tough discipline of the physical therapists who demanded we walk and walk in front of mirrors to the time of a metronome until we got our gait right. We used to joke that when we could walk, chew gum, and talk at the same time, we were ready to graduate.

One of my special heroes throughout this time was Sir Douglas Bader, the famous RAF fighter ace who had lost both legs in a training accident before World War II. "Tin Legs" Bader they called him. Colonel "Red" Reeder, who was assistant baseball coach at West Point, and who'd had his leg amputated as a result of wounds he suffered while serving in Normandy in 1944 with the 4th Infantry Division, was kind enough to send me a copy of *Reach for the Sky*, the excellent biography of Sir Douglas by Paul Brickhill, which I read and really enjoyed. Later, in the 1980s, while I was briefing the AirLand Battle concept to the British Army Staff in England, I had the opportunity to visit and chat with Sir Douglas before he died. He was a real inspiration for me.

The mid-1970s were the low-water mark for the modern American Army. Battered by its experience in Southeast Asia, and considered a third-rate force compared to the armies of countries like Israel, Great Britain, and even the former Soviet Union, it had to rebuild from scratch with a new, all-volunteer system of recruiting—offering incoming soldiers low pay and poor living conditions—while trying to bring a new family of weapons into production. A focus on possible NATO conflict with the Warsaw Pact dominated Army planning and procurement for the next twenty years. Out of this background there emerged a new theory of battle called "maneuver warfare." For Fred Franks, then a mid-level officer, it was a time to be in the middle of all the institutional growth and development at places like the Department of the Army, as well as the new organization called TRADOC.

Tom Clancy:	In the mid-1970s, the Army was going though some huge changes. Please talk about what you saw.
General Franks:	The Army and the nation had some extraordinary leaders in uniform in those days. Men of vision with the toughness to see those visions through. General Creighton Abrams. General Bill DePuy. They created the Training and Doctrine Command (TRADOC) and began the transition to an all-volunteer force. Also begun was the transition to a Total Army concept that involves completely the National Guard, U.S. Army Reserve, and active force components. So if we ever went to war again, we would have a tri-component combat team. A shared responsibility. That way the Administration, the Congress, and the American people would all have a stake in what happened to this new force. This, in a sense, reunited the Army and the American people—showing that this is "America's Army." To equip this force, Generals Abrams and DePuy authorized what came to be known as "the big five," a series of five new weapons systems that would provide the cornerstones for the Army in the 1980s. Included were the M1 Abrams tank, the Bradley fighting vehicle, the AH-64 Apache attack helicopter, the Blackhawk helicopter, and the SAM-D (Patriot) system.
Tom Clancy:	By the late 1970s, discussions of the new "maneuver warfare" doctrine were first published. Would you please talk about that for our readers?
General Franks:	The United States Army missed a whole modernization cycle during the Vietnam War. We were so focused on the execution of the war in Southeast Asia that we missed a materiel-upgrade cycle, as well as an intellectual and tactical development cycle. Thus, the speed and lethality of modern combat just leapfrogged over us during that intervening period. We were shocked into an awareness of this by the 1973 Arab-Israeli War. During that war there were more tanks lost than in the entire U.S. Army tank inventory in NATO. The Army needed to wake up to this, particularly if our central focus [i.e., vital strategic interest] lay in fighting and winning as a part of NATO in Europe. The question was, "What would this kind of combat be like for us?"

As it turned out, the '73 war proved to be an excellent model, a surrogate if you will. To get some idea of what was involved, General Bill DePuy, the TRADOC commander, and then-Major General Donn Starry from the Armor Center at Fort Knox, Kentucky, went over to visit the Golan and Sinai battlefields to see for themselves what we might have to face. The results of this visit were recorded in the doctrine that became the cornerstone of the 1976 release of the Army's basic field operations manual, FM 100-5. This was the first real break with our past doctrine. The new doctrine was initially nicknamed "the active defense," but really it was more than that. Much more than that. It was an acknowledgment of the speed and lethality of modern combat, as well as the Army's intellectual and doctrinal response to these changed conditions.

The Army continued to work on this doctrinal approach over the next few years, as the push to develop, test, and field the "big five" weapons started to bear fruit. Then we began to talk about the concepts of "maneuver warfare" and "operational level warfare," because the operational side of warfare was just not being practiced at the time. Our studies and wargames in the late 1970s were telling us that you could not win if you fought in a static or passive style of warfare. So a mobile, fluid style of warfare with depth was essential. Along with that, the idea of a "second echelon"—or what General Donn Starry (who was the principal author of the original concept) called "the extended battlefield"—started to be discussed. By the time of the 1982 version of FM100-5, the doctrine was given the more appropriate and correct name of "AirLand Battle." In this way, the Army simultaneously used the creation of new doctrine and the fielding of new weapons systems, as well as the creation of a full-time recruiting command emphasizing quality soldiers and early investments into a training doctrine that included the Army Training and Evaluation Program (ARTEP) and After Action Reviews (AARs) and what came to be known as the combat training centers [such as the National Training Center (NTC) at Fort Irwin, California], to bring it back to the forefront of the world's armies. By the time all this came to pass, we had experienced a virtual rebirth of the spirit and aggressiveness of the U.S. Army.

Tom Clancy: What was the bottom line on all of this high-level attention to the future vision and forces of the Army?

Tom Clancy:	Tell us about the first two days of moving into Iraq.
H. R. McMaster:	During the first two days of the campaign we covered short distances very rapidly, then halted for extended periods of time. Enemy resistance was very light and ineffective. Fox Troop made first contact with the enemy on the 24th. After a brief firefight, they captured large numbers of the enemy. Later that night we [Eagle Troop] detected a series of trench lines comprising an enemy infantry position. The Bradleys from 1st Platoon and my tank hit the position hard with 25mm fire, and TOW and tank HEAT rounds fired into the bunkers. The scouts also adjusted fire from the mortar section onto the area. The attack by fire drove the survivors toward Fox and Ghost Troops, who, after similar actions, took hundreds of the enemy prisoner. The next morning, as we moved further north, we encountered large numbers of surrendering enemy. We didn't have time to stop. So we ensured that they were unarmed and left them for units to the rear. . . . Many soldiers threw them food and water from the Bradleys, though.

Joe Sartiano's Ghost Troop had the only notable action in the squadron on the 25th. Ghost destroyed several MTLBs, captured several more, and drove them to the squadron command post. When we arrived for a meeting, the other troop commanders and I got to take them for a test drive. We knew that we had hit the security zone of the Republican Guard. The soldiers wore the Republican Guard insignia, and the unit had been well supplied and had new weapons and equipment. I told the troop not to assume that all the Iraqi units were as weak as those we encountered initially. I didn't want us to let our guard down. |

Eagle Troop
Command
Track   Lt. Hamilton
(XO – Lt.   Lt. DeStefano   Line of Advance
Gifford)   2nd Sqdn. TAC CP   Mortar Carriers   1st Scout Platoon

Meanwhile, the U.S. Army was not idle all this time. But the various designs for an IFV by American manufacturers had shortcomings that were not acceptable; and a number of prototypes failed to meet the Army's expectations. Part of the problem was cost. IFVs, with their expensive weapons mounts and sensors, cost two to three times what an APC like the M113 does. And the electronics necessary to control the various weapons systems simply did not exist in the 1960s. Yet with the development of solid-state integrated circuit technology in the 1970s, the time seemed right to try again.

By the early 1970s, the U.S. Army had made up its mind to get serious about IFV development, and contracted with FMC to produce something called the XM723 Mechanized Infantry Combat Vehicle (MICV). This was designed to accompany the new M1 Abrams tank into battle. The XM723 utilized a novel (and heavy) steel/aluminum hull armor, as well as components from the Marine Corps LTVP-7 armored amphibious tractor, to carry the weight of the heavier armor and a one-man 20mm cannon turret. At the same time, the Army was developing a new cavalry/scout vehicle, called the XM800, to replace the aging M114 cavalry vehicle. Congress, always cost-conscious, killed both programs, and ordered the Army to merge the two requirements into a common vehicle. From this, the M2/3 infantry/scout fighting vehicle was born. It was not a happy gestation. There were many compromises that went into the new vehicle—and many enemies to overcome.

The worst of these was a band of military reformers who inhabited various departments of the Pentagon in the late 1970s and early 1980s. Called by various names, such as the "lightweight fighter Mafia" or the "simple-is-better crowd," they advocated a return to simpler, low-technology weapons, bought in large numbers to meet the growing array of Soviet Bloc weapons systems coming on-line at the time. They despised the apparent complexity and cost of the weapons that all the services were then developing. And it was their sworn mission to find ways to kill these programs. Stories were leaked to the media. As

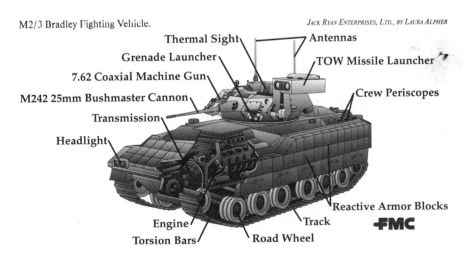

M2/3 Bradley Fighting Vehicle.

JACK RYAN ENTERPRISES, LTD., BY LAURA ALPHER

Thermal Sight
Antennas
Grenade Launcher
TOW Missile Launcher
7.62 Coaxial Machine Gun
M242 25mm Bushmaster Cannon
Crew Periscopes
Transmission
Headlight
Reactive Armor Blocks
FMC
Engine
Track
Torsion Bars
Road Wheel

an example, a scathing feature on the CBS program *60 Minutes* put the Bradley program under a cloud that would not go away until the 1991 war in the Persian Gulf. As for the people who count, the Bradley crews themselves, these guys love the vehicle and would hate to be in anything else. But enough about politics. Let's take a look at the inside of the Bradley.

The M2/3 Bradley—A Guided Tour

The M2/3 Bradley Fighting Vehicle (BFV) is a 30-ton/27.2-metric ton armored vehicle which carries a basic crew of three. Depending on which version of the vehicle you are looking at (the M2 is the infantry carrier and the M3 is the cavalry/scout vehicle), the Bradley can carry a variety of payloads. The M2 carries a small (six-man) infantry squad, while the M3 carries a pair of scouts, with additional radios, ammunition, and TOW missile rounds. In fact, the only noticeable external difference between the two versions of the Bradley are the side firing ports for the squad M16s, which are absent on the M3. The two rear-door firing ports are retained in both versions. And while the Bradley is nowhere near as sleek and low as its foreign counterparts, it does have the advantage of being the most heavily armed IFV in the world. The basic M2/3 variant which came into service in 1982 carried the same armament package as the M2A2s and M3A2s being produced today: a 25mm cannon, a 7.62mm coaxial machine gun, and a two-round TOW-2 ATGM launcher. In addition, the embarked troops or scouts carry their personal weapons, including the latest M16A2 assault rifle and the new AT-4 anti-tank rocket. The major differences between the earlier and present vehicles are in automotive performance and crew protection, plus an improved TOW missile system. There also is a better NBC breathing system, allowing the crew of the vehicle to breathe from a central hose-fed filter system, rather than using individual MOPP-IV (MOPP stands for Mission-Oriented Protective Posture) chemical-warfare suits.

The original M2/3 specification required armor capable of standing up to heavy machine-gun (.50-caliber/12.7mm) fire and fragments from artillery and mortar shells. But with the increasing use of heavier automatic cannons and man-portable anti-tank weapons, the armor on the newest Bradleys has been upgraded

The view through the rear ramp hatch of a Bradley fighting vehicle. The fold-down seats are evident, as well as the turret "basket" in the front of the vehicle. *John D. Gresham*

to survive hits from heavy 30mm cannons (like the GAU-8 on the A-10A "Warthog") and light anti-tank weapons like the Soviet RPG-7. The -A2 versions of the Bradley carry distinctive plates of add-on armor, plus deeper side skirts to protect the tracks and lower hull. Still, I must emphasize again that a BFV was never designed to stand up to heavy tank-gun or missile fire. The M2/3 was meant to work in a combined-arms team with the M1 Abrams, and not to fight enemy tanks by itself. (When this happened during the 1991 Gulf War, the Bradleys suffered in the process, but they gave as much punishment as they took.)

Like its heavy cousin, the M1, the Bradley has survivability features designed to save lives if something nasty penetrates the crew compartment or turret. To begin with, the M2/3 is equipped with a fire-suppression system that will snuff a catastrophic fire in the event of a fuel or ammunition hit. In addition, all of the M2A2s and M3A2s are equipped with spall liners to help keep fragments from injuring the crew if a projectile penetrates the hull armor. Unless a Bradley crew member is right in the path of a tank round or missile-warhead jet, he will usually survive. And if you are an enemy commander who sees a unit composed of M1s and Bradleys, you are going to shoot at the tanks first, since they are the greatest threat to your existence. This simple lesson from the Gulf War of 1991 has convinced the Army to attach M1s directly to cavalry squadrons in the heavy armored and mechanized divisions.

If you walk up to a Bradley, you enter the rear compartment where the "precious cargo" of personnel and ammunition is stored. You enter through an opening equipped with a heavy hydraulically powered ramp door. In the M2 versions of the Bradley, there are seats for six infantrymen and their weapons, plus stowage space for ammunition. The entire stowage layout of the IFV Bradley has recently been improved, with the 25mm ammunition stored under the floorboards and five TOW missiles in vertical racks. There are also firing ports and vision blocks for specially shortened versions of the M16 assault rifle—these have rarely been used in combat. The original IFV concept of infantrymen fighting from inside an armored vehicle just never has taken hold in the armies of the world. The simple truth is that in the end, soldiers have to get out in the open to use their individual weapons effectively. And all the smart bombs and guided missiles will never change that! The old grand master of science fiction, Robert Heinlein, was right. Someday, there will be starship troopers! The M3 cavalry/scout version of the Bradley is similar to the M2, except that four of the seats have been deleted to make room for more TOW rounds and the specialized radio gear used by the two dismount scouts that ride in the back.

As you move forward in the Bradley, you come to the turret basket. Inside the turret are seats and hatches for the vehicle commander (on the right) and the gunner (on the left). Both positions are surprisingly comfortable, even for folks over six feet. Much like the M1, there is a thermal sight boresighted to the gun tube, although the image appears on a red background instead of the more common green found on the M1. One quickly finds out that the most important thing in the entire turret is a socket wrench (the crews prefer a Sears Craftsman model, they tell us) used to manually operate the

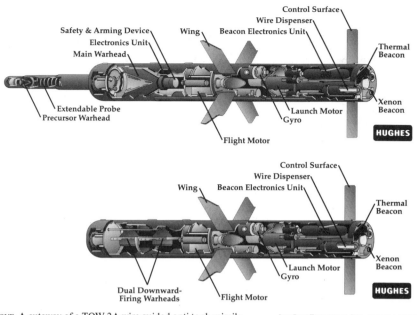

ABOVE: A cutaway of a TOW-2A wire-guided anti-tank missile.   JACK RYAN ENTERPRISES, LTD., BY LAURA ALPHER

BELOW: A cutaway of a TOW-2B wire-guided anti-tank missile. The two downward-firing warheads are located at the front of the missile.   JACK RYAN ENTERPRISES, LTD., BY LAURA ALPHER

machine gun and cannon feeds during reloading. This may seem rather crude for an armored vehicle costing over a million dollars a copy, but it is simple, it works, and it allows the ammo feeds to be kept compact (unlike, for example, the bulky and complex linear linkless feed system on the AH-64 Apache helicopter). The Bradley's gun is stabilized, and firing on the move requires very little correction. The cannon, machine gun, and TOW-2 missiles can be fired from either seat. Firing the weapons is not much more difficult than playing a video game. For the guns, it is simply a matter of selecting what kind of weapon and ammunition you want, using your hand grips to put the crosshairs on the target, and pressing the thumb trigger. Seventy-five rounds of 25mm armor-piercing ammunition and 225 rounds of high-explosive 25mm ammunition are usually loaded and ready to fire; two TOW-2 rounds and 800 rounds of 7.62mm machine-gun ammunition are ready-loaded as well.

The TOW missile is the primary anti-armor weapon of the Bradley, and it has a long and fascinating history. In 1958 the Army began to investigate the idea of a Tube-launched, Optically tracked, Wire-guided anti-tank missile; hence the acronym TOW. By 1963, Hughes had built a successful prototype, and the first production missiles were delivered to the Army in 1969. The first TOW was fired in combat by a UH-1 helicopter on May 22nd, 1972, knocking out a North Vietnamese T-54 tank at a range of 900 meters. Subsequent combat involving TOW provided evidence that it was an immediate classic. Tube-launching reduces the back-blast problem of earlier anti-tank weapons, for a small charge expels the missile from the tube. When ejected, the guidance fins pop out, and the rocket motor ignites at a safe distance from the launcher and its operator. Wire guidance means that the missile is under the gunner's control from just after firing until target impact, or until the wire runs out. On the TOW-2, there is a xenon beacon lamp in the tail of the missile, very bright in the infrared part of the spectrum, which makes it easy to track—even at night through fog, snow, or sandstorms, when the optional TAS-4 thermal sight is attached. All of this means that if the gunner keeps the optical tracker's crosshairs on the target, the missile will hit. System reliability in combat has been over 95%, which is amazing by any standard.

Mounted on a simple tripod, a complete TOW-2 system weighs 280 lb/127.3 kg and can be carried by a three-man crew. The minimum range of TOW is 65 meters, with a maximum range of 2.3 miles/3.75 kilometers (the length of the wire). The missile is subsonic: 623 knots/1,000 kph at peak velocity, which drops off rapidly after rocket-motor burnout. This happens shortly after launching, and then the missile just coasts along, with the TOW taking about fifteen to twenty seconds to reach maximum range. A twin TOW launcher is mounted on the Bradley Fighting Vehicle; and a single launcher can be fitted to a HMMWV, or almost any light wheeled or tracked vehicle. It is also used on the AH-1 Cobra and many other helicopters worldwide.

There have been five versions of the TOW, each with increasingly powerful warheads. Total TOW production is well over 500,000 units to date, with additional units still being produced as of early 1994. TOW is considered a "wooden round," and stays "fresh" in the canister, with a shelf life of up to twenty years in storage. All versions are compatible, as long as the launch-control unit has the correct software updates. The current version is the TOW-2B, which was developed to defeat the latest composite and explosive-reactive armor systems and is programmed to fly directly over the target, where a sophisticated sensor triggers two downward-firing warheads. These warheads have dense metallic liners that become "explosively forged projectiles" to penetrate the target's thin roof armor.

In 1993, the Army awarded Hughes a contract to produce 18,800 TOW-2A and TOW-2B missiles. This works out to a unit cost of about $9,800—a bargain price to kill a $250,000-dollar Soviet-type tank. To fire a TOW-2 missile the gunner simply keeps the crosshairs of the sight on the target, and the missile flies right into it. All you have to do is erect the two-round launcher (it can be reloaded from a hatch in the rear compartment roof), fire, and track the target in the sight. Despite fears that the new reactive appliqué armor on some tanks might defeat the TOW-2's warhead, experience in the Gulf War proved this to be false. TOW-2 can defeat any tank on the battlefield, with the possible exception of an M1 Abrams or a British Challenger II! The major limitation on use of the TOW system is that the Bradley has to be stopped to fire it.

The Bradley's guns are also easy to use. Although there is no laser rangefinder, range estimation is quite simple. You just punch in an estimated range on the weapons-control panel (there is a reticule in the sight to help you). Crews are trained to estimate range by judging the apparent height of the target in the gunsight—the closer the target the bigger it appears. Then you put the "death dot" of the sight onto the target and squeeze off a few ranging shots. Once you see the target bracketed, all you have to do is fire as many rounds as you need. The target quickly becomes obscured with dust and smoke from hits and near misses, and probably has also been destroyed as well. At 300 rounds per minute, it feels like you are shooting at the target with a fire hose! But gunners are trained to conserve ammunition by firing single shots until they see a hit on the target; then they fire three-round bursts until the target is destroyed.

Like the M1, the Bradley has a pair of the new SINCGARS radios for communications, though the current -A2 versions do not yet have access to the IVIS system. There is, however, a program under consideration for equipping the M2A2 and M3A2 with a downgraded IVIS terminal and GPS, but budget constraints may slow this process down until the new -A3 versions of the Bradley come on-line late in the decade.

If you exit the turret basket and head forward on the left side of the vehicle, you come to the driver's position and hatch. This is a very comfortable seat with a friendly instrument console. The Bradley is powered by a 600-horsepower turbocharged diesel engine with an automatic transmission. To start the vehicle no keys are required. You just pull a knob on the dashboard to activate the fuel pump, then push the shifter knob to the START position. The engine catches quickly and warms up smoothly. All that is then required is to shift down to the DRIVE position with the selector and press the accelerator. The vehicle accelerates quickly, and before you even know it, you are doing over 30 mph/48 kph. The Bradley has a top speed of over 45 mph/65 kph, and has no trouble keeping up with the M1 Abrams, either on-road or cross-country. The only trick to driving the -A2 version of the Bradley, which has an aircraft-style control wheel, is that it tends to be a bit loose in the rear during turns. However, you quickly get used to that; and after a while, driving the beast becomes a joy.

One chore that takes some time is setting up the Bradley for swimming. Because of all the rivers in Europe and Korea (where the Army expected to fight during the Cold War), the Bradley was designed to swim like its older cousin, the M113. But because of the extra weight of the turret and armor, the M2/3 requires a bit of additional buoyancy. To make this possible there is an erectable "swim curtain" of rubberized fabric over a steel frame to keep the upper deck free of water; and it works just fine.

In conclusion, the Bradley may not be the ultimate armored fighting vehicle, but it is a mature weapon system that is still evolving to keep pace with rapid changes in key technologies. Maybe not as sexy as an Abrams tank, but just as vital and necessary.

The M113 Armored Personnel Carrier

The M113 armored personnel carrier (APC) was the first modern "battle taxi" for infantry to use on the battlefield. It was designed to take advantage of technology breakthroughs of the 1950s in casting and welding aircraft-quality aluminum to create structures with the same strength as much heavier steel construction. This light weight allowed the FMC designers to use a relatively small automotive engine to drive a tracked vehicle with a substantial payload and the ability to float and swim across lakes, rivers, and streams. Even today when you look at one of the early-production M113s, they still have a clean, almost modern look. It is almost as if Frank Lloyd Wright took a shoe box and made it into a perfect shape to transport what the U.S. Army calls its "most precious cargo," the infantry.

Over the years, FMC and its licensees have produced something over 85,000 M113s for use by over two dozen countries (the Israelis call their M113s "Zeldas"). And while this venerable vehicle is no longer in production at FMC's San Jose, California, factory, it continues to be produced in places such as Turkey and Italy. Over 32,000 remain in service with the U.S. Army.

The basic model is the infantry squad carrier. This version of the M113 can carry a full squad of infantry (between ten and twelve soldiers), along with all of their weapons. It should be noted that this is a full squad, not the reduced unit carried by most IFVs like the M2 Bradley (which can carry only six dismount soldiers) or the BMP. It can also mount a machine gun (typically an M2 .50-caliber) or automatic grenade launcher (such as the 40mm Mk 19) on a rotatable mounting ("pintle") at the commander's hatch. To fire this external weapon, the commander has to open the hatch and stand on his seat. The troops sit in the back on bench seats with stowage for weapons and ammunition under the seats and on the sidewalls. They enter and exit the vehicle through a rear door equipped with a hydraulically powered ramp for easy access. The driver and gunner (if carried) sit in the front near the engine and transmission. Recently, FMC has begun shipping kits to improve the survivability of the M113 by installing interior spall liners, additional external armor, an upgraded engine and transmission, and external fuel tanks to reduce the risk of internal fires. Known as the M113A3, the conversion kit is being installed on existing M113s at several Army depots. These will be used by Army Reserve and National Guard units, and they will also be assigned for ambulance and support duties in active units. The M113A3 is a first-rate APC with many years of life ahead. It is cheap, rugged, and versatile. Not bad for a vehicle that was designed about the time I was a child.

While FMC and its licensees built tens of thousands of the basic squad carriers, they also built thousands of other M113 variants. The most important of these is the M577 command vehicle, which is used as a mobile command post by U.S. armored units. The M577 is essentially an M113 chassis with a raised roof and sides and additional generators to power the numerous radios that are stored in racks inside the rear compartment. The exact combination of radios depends on just what kind of command post it is (company, battalion, regiment, etc.) and whether it is stationed in a rear area or up front. It is additionally equipped with an expandable shelter tent in the rear of the vehicle to provide room for tables and map boards.

A 3rd ACR M113 configured as a fire-support team vehicle (FIST-V).
JOHN D. GRESHAM

It should be said that this venerable vehicle does have shortcomings, the most glaring being the inability of the M577 to operate effectively while on the move. Because it must be stationary for the onboard radio systems to operate, it becomes vulnerable to enemy artillery fire directed by radio direction-finding (DF) units. To help overcome this shortcoming, FMC is delivering kits to upgrade the M577 fleet to a configuration known as M577A3. This adds new radios and command and control displays, and provides some limited capability to operate on the move. As such, it will provide an interim C³ (command, control, and communication) capability until the introduction of the new XM4 C³ vehicle.

A more lethal version of the M113 is the M901. This model combines an M113 chassis with an erectable two-round launcher and sight (called a "hammerhead") for the TOW anti-tank missile. Designed to provide some anti-tank muscle to infantry units equipped with M113s, it represents a compromise between mobility and firepower (having TOW missiles only). A similar vehicle (called an M981 FIST-V), with a laser target designator in place of the missile launcher, is used by fire-support teams, though the Army is also starting to use M2/3s for FIST teams as well.

The versions of the M113 most likely to be found in armored units are the M106 and M125 mortar carriers. The M106 carries an 81mm mortar, while the M125 carries the more powerful 106mm mortar; otherwise the vehicles are identical. The mortars are mounted on a rotating base plate (much like a "lazy Susan") to fire through a large circular hatch in the roof. There is stowage for mortar rounds and radios for the various radio unit networks (nicknamed "nets"). The M106 is the most common version, with a pair of these usually assigned to each armored and mechanized company and cavalry troop. Their primary mission is to provide direct smoke and fire support, so that the captains and lieutenants who command small units can have their own artillery on call.

The M113 is an old soldier in comparison with the vehicles that replaced it. Around the world, various military forces have replaced the old aluminum

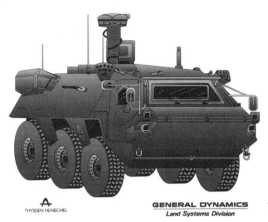

THYSSEN HENSCHEL

GENERAL DYNAMICS
Land Systems Division

The M93E1 Fox NBC Vehicle.
Jack Ryan Enterprises, Ltd., by Laura Alpher

box with such wonderful vehicles as the Bradley, the Warrior, and the BMP. But it should be remembered that no military force, including that of the U.S., can afford to base its entire armored force on expensive vehicle chassis like the Abrams and the Bradley. And so the M113 provides a good, simple base for a number of different applications; it is more than adequate for these jobs. There is an old saying, attributed to the former chief of the Soviet Navy Admiral Gorshkov: "Perfect is the enemy of good enough." For a good many of the jobs that the U.S. Army will have in the 21st century, the M113 will be more than good enough.

The Fox NBC Reconnaissance Vehicle

Of all the threats faced by the troops of the Allied Coalition during Desert Storm, the one that caused the most fear and concern was Saddam Hussein's arsenal of chemical and biological weapons. Built up over a period of years, the Iraqi stockpile was varied and battle-tested. In actual combat against the Iranians and the Iraqi Kurds, this array of chemical agents proved worthy of the popular nickname "the poor man's atomic bomb." And despite the claims that there were no known Iraqi chemical attacks during the war, the threat was very real and very frightening to the U.S. Army. Part of the reason for this fear was that the U.S. Army had, over a period of decades, allowed its ability to fight and survive on a chemical battlefield drop to near the bottom of its priority list. This is not to say that chemical and biological warfare did not figure into U.S. Army plans. The overpressure filtration system on the M1A1/2 Abrams tanks and the improved chemical-warfare suits issued to U.S. personnel in the Gulf were proof that some progress had been made. But these were passive measures only, and they did not solve several basic problems. For example, on a battlefield, you need to know *immediately* if a chemical attack has taken place. The U.S. Mk 8 detector, which was the Army's primary means of sniffing out chemical attacks before the Persian Gulf War, could begin to do that. That is, it could sound an alarm. But it could not determine how wide the contaminated area was, or what kind of agent had been used—two factors that are vital in allowing an army to fight effectively on a chemical battlefield. Prior to the deployment of U.S. forces to the Persian Gulf in 1990, the only way to evaluate the type and scope of a chemical attack was to send out soldiers with strips of indicator paper (specially treated to turn color in the presence of various chemical agents). Considering the variety and size of Saddam's chemical arsenal, something better was needed.

The Germans contributed the answer to this problem—an odd-looking armored vehicle called the Fuchs (Fox). Originally produced by Thyssen Henschel, the Fox (which has been type-classified as the M93) is a wheeled armored vehicle packed with special equipment for assessing the type and range of nuclear, biological, and chemical (NBC) attacks. Powered by a 320-hp diesel V-8 engine, and riding on six large rubber tires to maximize mobility, the Fox has a top speed of over 65 mph/106 kph and an unrefueled range of 500 miles/820 km. It carries a crew of four (driver, commander, and two sensor operators) in an armored citadel equipped with an advanced overpressure filtration/air-conditioning system. This allows the crew to operate in a fairly normal environment without the need for chemical suits and masks. And while the armor is not of the stature found in the M1A2 or the Bradley, it protects against machine-gun and mortar fire, which are major threats to reconnaissance vehicles such as the Fox. So comfortable is this system that many unit commanders I have met would like to take over a Fox for their personal command vehicle. One went so far as to compare the ride, seats, and air-conditioning to a really nice GM or Winnebago motor home!

The real payload of the Fox is the integrated sensor suite located in the rear compartment of the vehicle. The most impressive of these is a funny-looking system used to survey and assess areas suspected of being contaminated—or "slimed"—by chemical agents. On the rear of the vehicle are a pair of mechanical arms carrying rollers with a sticky silicone rubber coating. One at a time, these are lowered to the ground and allowed to roll along the surface, picking up traces of any chemicals that might be present. Once this has been done for a short time, the arm is raised (and the other lowered) to a sensor head for analysis by a mass spectrometer. This has the capability to identify dozens of chemical agents, from mustard gas to the latest lethal nerve agents. The computerized analysis gear is tied to an inertial navigation system (similar to the POS/NAV system of the M1A2) which maintains a map record of the area that has been surveyed. In addition, there is a sensor to detect radioactive fallout or nuclear contamination, if present. Once an area has been surveyed, the data can be relayed back to the commander of the friendly forces via a pair of SINCGARS radios. And should it be required, the Fox can lay a trail of brightly colored markers, to flag a contaminated area or show a safe transit lane through such a zone.

Sixty of these sly little vehicles were donated by the Germans to the U.S. to support the coalition forces in the Gulf. Ten went to the two Marine Corps divisions on the eastern end of the front, and the other fifty went to Army units in the VII Corps and XVIII Airborne Corps. Two were assigned to each brigade or regimental headquarters to scout for zones of chemical contamination during the drive into Iraq and Kuwait. And while you did not see much of them on CNN or the other news services, rest assured that right behind the leading armored wedges of the U.S. forces were the Foxes doing their job. Fortunately, it seems there were no actual chemical attacks, though a number of Iraqi storage bunkers were shattered by bombs and artillery shells, causing chemical spills and downwind diffusion of toxic agents. The work of the small force of Foxes continued after the cease-fire; they were busy for weeks surveying the battlefields of Iraq and Kuwait, looking for traces of deadly chemicals. By all accounts, they were extremely successful, with not one known Allied casualty due to Iraqi chemical weapons.

The positive experience that U.S. forces had with the Fox has translated into a program to acquire more of the vehicles and upgrade them with new and improved capabilities. General Dynamics Land Systems (the maker of the M1A2 Abrams), in partnership with Thyssen Henschel, was selected by the Army to provide additional Fox NBC reconnaissance vehicles. In addition to the sixty vehicles already donated by the Germans, another forty-eight Fox/M93 vehicles have been procured to support other units that the U.S. military might send into combat. In addition, the entire fleet of Foxes will be upgraded with GPS receivers to improve navigational accuracy. Planned enhancements include additional sensors, a central computer with commander's display and printer, as well as an erectable mast capable of sensing chemical gas clouds and their characteristics (size, agent, speed and direction of movement, etc.). By the time this procurement has been completed later in the 1990s, the improved Fox/M93A1 vehicles of the U.S. Army will be the most capable NBC reconnaissance vehicles in the world. The only thing that Fox will not have is numbers, for the Army could probably use two or three hundred of the handy little vehicles. Nevertheless, the hundred or so assigned to field units will help the Army cope with the NBC threat on future battlefields, as well as with the growing possibility of toxic or radioactive terrorism. With periodic updates of the mission software and equipment, the NBC Foxes will continue in this role for several decades to come. *Danke schön, Deutschland!*

Other U.S. Armored Vehicles

Thus far, we have only discussed a few of the vehicles that make up the armored forces of the U.S. Army—those with an obvious military function. The next few vehicles we will explore are more like what we might see on our local streets. What makes these different from their civilian counterparts is their need to do their jobs on the battlefield. This makes them more expensive than their civilian equivalents, and sometimes compromises their performance by comparison. But just as much as the M1A2 and the Bradley, they are combat vehicles.

M88 Armored Recovery Vehicle

Fact: Tanks and other armored vehicles are heavy. Another fact: Tracked and armored vehicles break down, sometimes a lot! And when a big iron beast like an M1A2 breaks a track or its transmission dies (and these things do happen), then you need one hell of a tow truck to drag it back to the repair yard and get it going again. And this is the job of the M88 armored recovery vehicle. Built by BMY Combat Systems in York, Pennsylvania, the M88 is a heavy tracked armored vehicle equipped with the necessary tools to extract or tow a heavy armored vehicle back to a field-maintenance unit that can repair and return it to service. Even vehicles that have been heavily shot up can frequently be put back into service with just a few hours in the shop. For example, during Desert Storm, if an M1 suffered turret damage, it could be taken to a repair line, where the whole turret could be replaced in just a few hours. The problem is to get it there.

As we noted earlier, the current version of the M88, known as the M88A1, is somewhat underpowered (only 750 hp) and light (56 tons/50,909 kg) compared to the M1. This makes it difficult to tow one of the big Abrams tanks if it is heavily damaged or stuck. To make up for these shortcomings, the Army and BMY have developed and tested the M88A2. It weighs 70 tons, and is powered by a 1,050-hp turbocharged diesel engine and beefed-up transmission, giving it the capability to recover even the largest armored vehicles. The vehicle's boom/hoist winch has a 70-ton pull; it can upright an overturned tank or remove the 25-ton turret of an M1A2. Deliveries of this new version should begin in 1995, though subsequent deliveries may be delayed by funding. Nevertheless, the Army wants this vehicle very badly! Because they never want another M1 having to fight for its life waiting for a tow!

M9 Armored Combat Earthmover (ACE)

Is there anything in the world more "blue collar" than a bulldozer? For most of this century, this distinctly American construction implement has been changing the face of the earth. It's also a significant piece of combat equipment. Bulldozers have been used by American combat engineering units since they were introduced in World War II by the famous U.S. Navy Construction Battalions known as "Seabees." Since that time, they have been used by all four branches of the service in everything from base civil-engineering projects to building defensive positions in the field.

Armor commanders have always wanted a bulldozer that could accompany them into combat. An armored bulldozer resistant to small-arms fire and shell fragments could dig fighting positions and breech obstacles while under fire. But the project to design such a bulldozer was way down on the Army's priority list as compared to projects like the Abrams MBT and the Bradley IFV. In addition, the original contractor for the project, now called the M9 Armored Combat Earthmover (ACE), fell behind schedule and went over budget. In the end, after accepting the prototypes and conducting operational trials, the Army reopened the contract for bids and awarded it to BMY. BMY got the vehicle into production; and ACEs began to roll off the line in 1988.

The M9 ACE looks like a metal box with small tracks on the sides and a bulldozer blade on the front. Inside is an armored one-man cab, with its own NBC filtration system to allow the driver to operate the ACE in a contaminated environment. The M9 is powered by a 295-hp Cummins V903 diesel engine running through an eight-speed (six forward, two reverse) automatic transmission. It can move at up to 30 mph/49 kph on roads, so it can self-deploy, without requiring a low-boy tractor trailer to haul it up to the front line. It can even swim at 3 mph/5 kph. It is air-transportable by any of the major transport aircraft from the C-130 Hercules to the C-5 Galaxy. And thanks to a unique hydropneumatic suspension, it can scoop up 8.7 cubic yards of earth in its front scraper bowl in a single push, and can perform other earthmoving tasks such as ditching, scraping, hauling, winching, and towing.

But most important, the M9, and a similar British vehicle called the Combat Engineer Tractor (CET), are the only earthmovers capable of working and surviving under fire. Tanks can be fitted with crude dozer blades, to plow through obstacles, but tank drivers are not trained for earthmoving and construction work. For that, you need an ACE.

Enter Saddam Hussein and his decision to roll into Kuwait in August of 1990. Almost as soon as the first units of the 82nd Airborne Division and the Marines began to land in northern Saudi Arabia, their commanders were screaming for equipment to help them dig in. The fear was that with only light units and limited weapons, these forces would be nothing more than "speed bumps" if Saddam's heavy armored forces were to move south. Historically, forces that are properly dug in are three to five times more effective than those out in the open. Thus, within days of the order starting Desert Shield, BMY received a call to move all the M9s that could be collected from the delivery holding area and production line (about ninety-nine, I am told) to the airbase at Dover, Delaware, to be shipped by C-5 Galaxy transports directly to the U.S. forces in Saudi Arabia. The only preparation for their new duty was a coat of new desert-tan CARC (Chemical Agent Resistant Coating) paint to help them blend into the Arabian terrain. So vital were the ACEs that they went ahead of such important equipment as AH-64A Apache attack helicopters, and even ahead of ammunition! As soon as they arrived, they were trucked to the forward units that needed them so dearly. For the next few months, they were constantly busy digging fighting positions, anti-tank ditches, and a variety of other earthworks. Even the Marines out near the coast asked for some ACEs, and had thirty delivered in short order.

Later during Desert Shield, when the threat of an Iraqi invasion of Saudi Arabia had been blunted and the idea of an offensive to liberate Kuwait had taken hold, it was the M9 and other engineering tools that would make it possible. Equipment like the ACE allowed the U.S. to plan for breaching the defenses that Saddam's forces had constructed along the border with Saudi Arabia. Trials with M9s told the U.S. Army that the ACE was capable of rapidly breaching the tall sand berms that the Iraqis had built along the border. The challenge here was for each M9 to rapidly cut three to five lanes through the berms (which were 20 to 30 feet/6 to 9 meters tall, and 50 to 80 feet/15 to 25 meters thick) so that the armored spearheads could rapidly transit into the enemy positions beyond. And so when Desert Storm opened for the ground forces in February of 1991, the M9s led the way. As soon as the berms had been cut, the ACEs followed behind the armored spearheads, ready to do any job that might be required of them. It was a performance that was almost flawless, and almost unknown to the rest of the world. But just ask the Army captains and lieutenants who led those forward platoons and companies/troops whether or not they were glad the M9s were there. I think you will find them unanimous in their praise.

U.S. Army Trucks and Other Transport Vehicles

The U.S. Army Truck/Transport Philosophy

While the German Wehrmacht introduced the world to their Blitzkrieg concept of mobile warfare, it is important to remember that even after two years of war, the German Army was still primarily a horse-drawn force. In fact, it was the Americans who created the world's first truly mobile army. The key to this mobility was wheeled vehicles. Virtually every American serviceman knew time in a jeep or heavy truck. Even the famous 82nd and 101st Airborne Divisions rode into the Battle of the Bulge in late 1944 in trucks, not airplanes and gliders. Wheeled military vehicles are not sexy. They are not flashy. They are simply necessary. For without the "low-boy" trucks to carry armored fighting vehicles to the front (they wear out tracks and transmissions relatively quickly), fuel trucks to carry diesel to fill their tanks, and cargo trucks to carry food, spare parts, water, and all the other stuff needed to make war, there would be no armored warfare.

The current generation of Army wheeled vehicles was designed with the idea of moving with and supporting highly mobile tracked vehicles like the M1 Abrams MBT and the M2/3 Bradley fighting vehicle. They can climb the same hills, ford the same streams, and move across the same kinds of terrain as their heavier counterparts. This is why the TV pictures of the armored wedges of American vehicles showed so many trucks and other wheeled vehicles moving in mixed formations. They move with the armor, because they were designed to. This ability doesn't just happen. It is designed in from the start, and represents a commitment of decades to the concept that all of the Army's vehicles should be able to move together.

A 3rd ACR M998 HMMWV
"Hummer" in its natural environs.
JOHN D. GRESHAM

High-Mobility Multipurpose Wheeled Vehicle (HMMWV)

Program History—Every war tends to create its own icons. Who can forget the F-4 Phantom jet, the UH-1 "Huey" helicopter, or the AK-47 assault rifle from Vietnam? In Desert Storm, along with the F-117A stealth fighter, the SCUD surface-to-surface missile, and the Patriot anti-aircraft missile, there was the High-Mobility Multipurpose Wheeled Vehicle (HMMWV, pronounced "Humvee") or "Hummer" (AM General's trademarked product name). Virtually every serviceman in Southwest Asia rode one sometime during Desert Shield and Desert Storm. President and Mrs. Bush traveled in one, and even ate their Thanksgiving dinner on the hood. And while the cutting edge of the Army in Desert Storm may have been concentrated in a few thousand M1 Abrams tanks and M2/3 Bradley fighting vehicles, the numbers of HMMWVs used during the war was about twenty thousand! Used for everything from a personal transport for officers to a medium truck mounting TOW anti-tank and Stinger anti-aircraft missiles, these served with a distinction worthy of their distinguished grandfather, the World War II Willys Jeep. The HMMWV is the replacement for the Army's last jeep model, the M151 (built by Ford); and today it is the most widely used vehicle in the U.S. military.

So just what is a HMMWV? The U.S. Army calls it a four-wheeled vehicle in the "medium" (4,000-to-10,000-lb/1,818-to-4,545-kg) class. But the forces that use the HMMWV (all four branches of the U.S. military, along with numerous foreign countries) look upon it as a "do everything" vehicle. It performs all of the missions that used to be accomplished by the old M151 jeep, as well as the old 1-1/4-ton truck (called a "five quarter") and six other truck types. This simplifies the skills needed to operate and maintain it, and greatly reduces the need for separate lines of spare parts.

Thanks to TACOM's unified-design philosophy, the HMMWV is probably the most mobile and rugged wheeled vehicle ever built. Like its more deadly tank and IFV counterparts, it was designed to the same standards of mobility as all the Army's vehicles. It climbs the same hills, fords the same streams, and has the same kind of redundancy and reliability as every vehicle of the current generation. That is the key to the Hummer's extraordinary success.

The basic version is known as the M998, which provides the chassis for a whole family of different variants. Some of these include:

- The M998 open-topped cargo/troop carrier used for general purpose operations and transport.
- The M1026 armament carrier capable of mounting an M2 .50-caliber heavy machine gun, an M60 7.62mm light machine gun, or an Mk-19 40mm grenade launcher.
- The M966 TOW-2 missile carrier. It utilizes a light TOW launcher tripod (which can be dismounted), and has room for six TOW-2 rounds under the rear roof shell.
- The M966 mini-ambulance, and the M997 maxi-ambulance. These are the primary medevac ground assets of the U.S. Army, and proved to be extremely effective during Desert Storm.
- The M1037 shelter carrier, which is used for a variety of functions from command and control to mobile workshops.
- The M1097 "Heavy Hummer," which may be armored against small-arms fire. These 10,000-lb/4,545.5-kg vehicles are being adapted as a squad carrier for light infantry, and as a new scout vehicle being developed by TACOM.

There can be little doubt that AM General and the Army will be developing additional HMMWV variants. They should be successful, for AM General has a successful history going back over fifty years.

The story of the HMMWV is an evolutionary one. It is produced by the same company, AM General (a direct descendent of Willys), that produced the original Jeeps. And while AM General has changed ownership more often than just about any other defense contractor (from Kaiser Jeep, to American Motors, to LTV, and finally to the Renco Group), the HMMWVs that come off of the line today are being made by the sons and daughters, and even the grandsons and granddaughters, of the men and women who built military vehicles during World War II.

It would be nice to say that the HMMWV has the same kind of high-tech, sexy design philosophy as the M1 or Bradley. In fact, the design of the HMMWV is a marvel of engineering conservatism. The primary chassis structure of the Hummer is a pair of massive steel beams that run the entire length of the vehicle. The HMMWV rides on four specially designed wheels, each driven by a novel geared hub. The bodywork is almost entirely built of lightweight aircraft-quality aluminum alloy, so resistant to corrosion that AM General warrants the HMMWV against it for fifteen years! The Hummer is powered by an eight-cylinder, 150-hp General Motors diesel engine driving a three-speed automatic transmission. The result of all this is a vehicle with an extremely low center of gravity, which can climb almost any hill, ford some 2.5 feet (76 cm) of water, run across ice, snow, and sand with ease, and climb steps and logs almost two feet tall.

Driving the Hummer—When you first walk up to a Hummer, the first thing that strikes you is just how incredibly boxy it is. AM General is proud of saying that not one penny of taxpayer money has ever gone into making the HMMWV stylish or attractive. In fact, it reminds one of the Air Force's A-10A Thunderbolt II (known affectionately as the "Warthog"). Both are designed to be functional, not pretty. But to the folks at AM General and in the Army it is a thing of beauty. "Functionally elegant" is how they describe it.

The military HMMWV has little in the way of creature comforts, and the back seats of early models of the HMMWV are considered a form of torture by some passengers. If you are looking for luxury, forget this beast. As you sit down, it is vital to buckle up the safety belts. The ride of a HMMWV can be rough when crossing broken terrain, and most military HMMWVs have no doors! Only the weapon carriers and a couple of other variants have hard doors, though all other HMMWVs have the provision for canvas doors and covers. Each military Hummer is covered in the same kind of chemical-resistant CARC paint that is used on virtually every other Army vehicle being produced today. In most cases there are mountings for radios between the driver's and front passenger seats, and many HMMWVs have dashboard mounts for the military Trimble GPS receiver.

As you view the dashboard, you will be decidedly underwhelmed if you are looking for the state-of-the-art electronics of the M1A2. There are about the same instruments you might expect to find in an old Volkswagen: speedometer, fuel gauge, and a few indicator lights. The steering wheel is bare black plastic, with a pair of shifter knobs, one for selecting the normal gears (NEUTRAL, DRIVE, REVERSE, etc.), and a second one for selecting high or low range for the transfer case. As with most military vehicles, there is no key, just a switch on the upper left portion of the dashboard with positions for START, OFF, and DRIVE. To secure the vehicle you pull a steel cable out from the dash and padlock it to the steering wheel. Starting the Hummer is quite straightforward, and the engine quickly comes to life. For those used to the soft purr of a gasoline engine in a normal road car, the decided rumble of the HMMWV's GM diesel may seem rather extreme. Actually, the diesel in the Hummer is quite smooth, with a very nice power curve. The key to getting the most from the powerful diesel is to think a little ahead of where you normally would driving a conventional road car: There is a slight throttle lag after you push down the accelerator pedal before the power finally begins to come on.

Unlike many off-road vehicles, the HMMWV is always in four-wheel drive (unless you put it in neutral for towing). You choose the high- or low-gear range on the transfer case. You would choose HIGH for highway driving. Surprisingly, driving the HMMWV requires a delicate touch, and not the rough kind of driving usually associated with conventional trucks and four-wheel-drive (4WD) vehicles. In feel, it is something like a Cadillac road car, with gentle steering commands and thoughtful use of power required to get the most out of it. And yet in all other respects, it does not drive like a luxury car. The best word to describe the sensation of driving the Hummer is authority. Whether it is climbing over rocks (the HMMWV has a minimum of 16"/41 cm of ground clearance) or fording streams (up to 30"/76 cm of running water), it feels capable of accomplishing any task. Of course the HMMWV can't be driven recklessly. Rather, it is a vehicle which rewards a thoughtful, careful driver with performance that has to be seen to be believed. A major at the National Training Center at Fort Irwin, California, once told me that if you ever run into a piece of terrain that a HMMWV cannot reach or climb, you probably don't want to own it anyway!

The Hummer is capable of several extraordinary tricks. For example, there is the matter of climbing steep hills, and then descending them! To climb almost any hill you might want to get to the top of, all you have to do is tighten your seat belt (so you don't have to worry about falling out), drop the shifter down into the D1 position (low range, low gear), and gently apply the accelerator. Almost like magic, you will head up the hill at a steady, albeit slow, rate. The key here is applying steady, constant power to the fat tires. (Incredible as it may sound, mud, gravel, soft sand, and snow can be dealt with simply by slowing down a bit and gently applying the power.) As amazed as you will be to get to the top of a hill, getting down is even more impressive. The trick is to again put the transmission into D1 position, and head on down. The major point is to make sure that you never touch the brakes! Odd as this may sound, when descending, you are actually just hanging on the compression of the HMMWV's diesel engine, and touching the brakes would just lock them up and might start you sliding dangerously. It is an amazing feeling to look down a slope that you frankly doubt you could climb on foot, and yet you are driving down it with a feeling of authority and confidence. This was demonstrated most convincingly by a major of the Dragon Observer/Controller team up on the live-fire range at the NTC. When he wanted to show some members of my research team the range instrumentation at the top of a mountain (a particularly tall and treacherous-looking rock pile), he just shoved his Hummer into D1 and headed up without giving it a thought. The "path" (if you care to call it that) was just gravel and rock shale. He later told us that the only ways to the top were via helicopter—and HMMWV!

I'll end my discussion of the Hummer for now by describing a particularly famous photo from the Persian Gulf War (the war was filled with Hummer stories). It shows a HMMWV of the 82nd Airborne Division loaded almost to the breaking point (the wheels and axles are just jammed all the way up into the wheel wells) with ammunition and equipment, carrying a couple of airborne troopers, and towing a 105mm howitzer to boot! Later we will describe the Army's plans for this fascinating vehicle, which has in fact become so popular that AM General sells commercial and private versions of the HMMWV. And while they are not cheap ($40,000 to $60,000 as of 1994), they are probably the finest 4WD/utility vehicles being built today. I should know since I've bought one for myself!

M1070/M1000 Heavy-Equipment Transporter System (HETS)

One does not just unload a battalion of M1A2s or Bradleys and run them into combat a couple of hundred miles to the front. Armored vehicles are mechanically demanding brutes which generate a lot of wear and tear on themselves for every mile they run. Thus, since it became common for armies to use heavy armored vehicles, there have been special transporters to carry them to the fighting fronts. Resembling low-boy semi-tractor-trailer rigs, these are designed so that the armored vehicle drives up a ramp onto the trailer, which then hauls it to a forward assembly area. The problem with the current HETS, known as the M911/M746 with 747 (the first number is the designator for the tractor, the second is for the semi-trailer), is that the greatly increased weight of the M1 MBT limits it to paved roads and speeds no faster than 15 mph/25 kph.

To make up for these shortcomings, the Army is starting to field a new unit, the M1017/M1000. Produced by Oshkosh Corporation of Oshkosh, Wisconsin, it can move both on and off road with a seventy-ton payload (such as a loaded Abrams), at speeds approaching 30 mph/50 kph. And in perhaps its most important role, it will be able to maneuver on the Army's new class of roll-on/roll-off transport ships without even having to use reverse gear. This allows the M1017/M1000s to move their vital cargo off the ships with a minimum of support equipment and dock facilities. And that can be the difference between winning a battle, and having the bad guys sunning themselves on the beaches of one of our allies.

M939 5-Ton Truck

Army units need a lot of stuff. Stuff to eat. Stuff to drink. Stuff to shoot. And all this stuff has to be moved somehow. This means trucks. Lots of them. The most common of these trucks is the classic 5-ton (this refers to the payload) M939. The newest version of this venerable design, the M939A2 is produced by the BMY Truck Company. The unique feature of the M939A2 is a new central tire-inflation system that allows the driver to alter the tire pressure, and thus the traction, with the flick of a switch from inside the cab. This improves the performance of the M939A2 in soft sand and mud.

Heavy Expanded Mobility Tactical Truck (HEMTT) Family

The Heavy Expanded Mobility Tactical Truck (HEMTT) family of trucks is designed to provide forward units with fuel, water, dry foods, and other supplies. Produced by Oshkosh Corporation, they also provide a platform for cranes and other service equipment needed to maintain other Army vehicles. With a 10-ton/9,090-kg payload and powered by a 445-hp diesel engine, the HEMTT can move at 55 mph/90 kph across a variety of terrain. There are currently five configurations, all of which gained combat experience during Desert Storm. The M977 is a light cargo carrier with a crane; the M978 is a fuel tanker with a 2,500-gallon/9,433-liter capacity; the M983 and M984 are tractor and wrecker versions respectively; and the M985 is a heavy cargo carrier with a material-handling crane. All in all, the HEMTTs are a capable series of transporters, with a great deal of growth left in the basic design. It is almost certain that the Army will be developing further variants of this excellent vehicle.

Coming Developments

In the 1980s, with the introduction of the AirLand Battle Doctrine and the institution of synchronized-maneuver warfare, the U.S. Army underwent a major revolution. And just when the bad guys are dreaming up more trouble, here comes General Gordon Sullivan, the Army Chief of Staff, with some bad news. Because the revolution of the 1980s worked so well, the Army has decided to have another one in the 1990s. By 1997, virtually all of the systems described below will be on-line and in actual maneuver units. Some of them, like the M2A3/M3A3, are upgrades of existing systems. Others, like the XM8 Armored Gun System (a powerful light tank), represent entirely new ideas. These new systems are designed to give excellent value for the limited procurement dollars that will be available in the late 1990s. And for those of you who consider defense spending bloated, consider this: Since the close of Desert Storm in 1991, the Army procurement budget (money to buy new equipment from tanks to desktop computers) has dropped by two-thirds! In spite of this, General Sullivan is still pursuing his idea of digitizing the battlefield by 1997 with passion and vision. And the way he has been going, it might just work.

The M2A3/M3A3 and Bradley Stinger Variants

By now you might be wondering when the Bradley will get the same digital makeover as the M1A2. With its onboard digital data bus and the IVIS vehicle network system, the M1A2 enjoys a tremendous command and control advantage over just about everything else on the battlefield. But the current version of the Bradley has none of these systems; it rides beside the new tank dumb and blind by comparison. Without an IVIS interface, the best Bradley commanders can do is listen for voice radio messages from their Abrams companions.

FMC is leading the effort to get some type of digital command and control system for the M2A2/M3A2. And GDLS modified some six M2A2s into

IVIS-capable units for testing and evaluation at Fort Hood and the NTC. In the short term, there will be a program to upgrade the Bradley -A2 fleet with interim kits that can be installed at Army depots or bases. This will probably include the addition of a GPS receiver, a laser rangefinder, and perhaps a simple terminal for the vehicle commander. It will probably not be tied to any of the vehicle systems, and will simply feed its data through the SINCGARS radios. Thus, such vehicles would be IVIS-capable, but not really fully integrated.

Another version coming soon is the Bradley Stinger Vehicle. This variant of the Bradley is meant to fill some of the gaps in the Army's air defense. Despite excellent performance by Army units firing the Patriot anti-aircraft missile during Desert Storm, things might have been quite different had Iraq chosen to attack forward units with their remaining ground-attack aircraft and helicopters. This is because the forward air defenses of the Army maneuver units were limited to a few obsolete systems, such as the Chaparral anti-aircraft missile, the PIVADS 20mm anti-aircraft gun (an M113 carrying a Vulcan rotary cannon), and the Stinger-RMP light anti-aircraft missile. The Stinger is the best of these, with a state-of-the-art seeker head which can see through almost any combination of decoys and jamming. The problem for the U.S. Army's armored forces is that Stinger is currently deployed only with man-portable teams and a handful of Avenger anti-aircraft vehicles (HMMWVs with a pedestal mount for eight Stingers and a .50-caliber machine gun).

To help rectify this problem, the Army is developing a version of the Bradley designed as an anti-aircraft system to defend the armored forces. Currently in test, the Stinger Bradley has a new turret and extra stowage in the rear compartment for missiles and 25mm gun rounds. The turret has been modified with the TOW-2 launcher replaced by a four-round launcher pack of Stinger missiles. A new command and control system allows targeting data to be fed from the Forward Area Air Defense System (FAADS) early warning net. The fire-control system was modified to allow firing of the Stingers and engagement of fast-moving air targets with the 25mm chain gun. This also requires Identification Friend or Foe (IFF) interrogation gear. The exact number to be produced, as well as scheduling and fielding doctrine, is still being considered at TRADOC; but the Bradley Stinger is on its way to becoming the best mobile anti-aircraft platform the Army has ever deployed. The only problem on the horizon is that old bugaboo of money. Somehow the Army must find a way to shoehorn the Bradley Stinger program into an already stretched budget. They probably will; they know that not every future enemy will be as ineffective as the Iraqis!

A long-term improvement to the basic Bradleys will be the M2A3/M3A3 Bradley fighting vehicle. Under this program, FMC would take existing M2/3 chassis from the early Bradley production runs, strip them down, and remanufacture a new vehicle from them, much like the U.S. Army M1A2 program at GDLS. The vehicles would be fully integrated through an MIL STD 1553 data bus, with all analog systems converted to more reliable digital ones. Some of the ideas under consideration include an independent thermal viewer for the commander (like the CITV on the M1A2) and a laser rangefinder/designator to support scouting and laser-guided weapon delivery. New thermal sights would probably take advantage of the second-generation FLIR technology used in systems like the RAH-66 Comanche scout/attack helicopter, providing much better display resolution than existing systems. There will also probably be an inter-vehicular network compatible with the IVIS system on the M1A2. This could be a second generation system that overcomes the data bottlenecks of the current IVIS system (which has a transfer rate of between 1,200 and 2,400 bits per second, about the speed of a relatively slow computer modem). On a more mundane note, FMC is experimenting with a set of inflatable floats to replace the present swim curtains. There are even some unconfirmed rumors of work on plastic fighting vehicles at FMC. Layers of plastic, Kevlar, and other composites might replace portions of the ballistic hull armor on vehicles such as the Bradley. Such a vehicle would have the advantage of being lighter, cheaper to operate (better fuel economy, less wear and tear on components), and might even be "stealthy," due to the lower radar cross section (RCS) of a plastic upper hull.

For now such dreams will remain in the design shops of FMC and the halls of TACOM in Warren, Michigan. Money for new systems is very scarce right now, and the effort to design and field the -A3 Bradley will have to wait its turn in the funding cycle. And more mature programs, such as the M1A2 and the AH-64 C/D Longbow Apache, which are in production or testing, have a higher priority. Nevertheless, sometime around the dawn of the 21st century, watch for a new version of the Bradley to take its place among the newest digital systems fielded by the U.S. Army.

The XM8 Armored Gun System

Quietly tucked into a corner of the manufacturing floor at FMC's San Jose, California, plant is an assembly line that may be soon producing one of

XM8 Armored Gun System. Note the driver's hatch with panoramic vision blocks. The rather narrow tracks are indicative of the vehicle's light weight. *FMC CORPORATION*

the most exciting armored vehicles in history, the XM8 Armored Gun System (AGS). Actually a highly capable light tank (that term having fallen out of favor within the U.S. Army), the XM8 is a response to the problem of light cavalry, infantry, and airborne forces lacking an armored punch.

The idea of a light tank is that it can destroy anything weaker and hide or run away from anything stronger. At less than half the weight and half the cost of a main battle tank, but carrying a nearly equivalent weapon, a light tank theoretically gives you more bang for the buck. In practice, light tanks have usually proven to be more expensive, less effective, and less survivable than their designers hoped (which is why the term "light tank" is out of favor).

Since the Vietnam War, armored fire support for light forces has been the job of the M551 Sheridan light tank. Unfortunately, the M551 never lived up to expectations, and has been a disappointment to its users. Its aluminum armor proved ineffective against hand-held anti-tank weapons in Vietnam, and its complex 152mm gun/missile launcher was chronically unreliable. It also had a nasty habit of shedding a track in tight turns. In fact, the only really quality service the Sheridan has been able to give the Army has been as a cheap and available chassis for the OPFOR teams who simulate enemy tank battalions at the NTC at Fort Irwin, California. Today (early 1994), the last active combat unit equipped with the M551 is the airborne tank battalion of the 82nd Airborne Division. And the only reason it remains is that the Sheridans can (just barely) be airdropped. Other than that, the Army would just as soon scrap them.

But the failure of one weapons system in no way eliminates the need and mission for light armored support to light infantry, cavalry, and airborne troops. In fact, the need has grown with the loss of basing rights for American forces (such as in the Philippines) in the 1980s and 1990s. And while General Sullivan plans to keep eight full combat brigades' worth of equipment on ships ready to steam, there still will be a need to airlift rapid-response forces into a crisis area. To fill this need, the Army initiated the AGS program in the 1980s, to provide a light, survivable armored vehicle with good firepower that could be air-transported almost anywhere. A competition was held, and FMC was declared the winner. Six XM8 prototypes will be completed by the fall of 1994. If the program stays on schedule, the first unit of XM8s, an airborne armored battalion of the 82nd Airborne Division at Fort Bragg, North Carolina, will come on-line in the fall of 1997.

The XM8 has an aluminum hull and turret, and is armed with a soft-recoil 105mm gun capable of firing the same ammunition as the early versions of the M1 Abrams. The Army has adapted an autoloader from the U.S. Navy's 5" Mk-45 gun mount to handle and load the shells in the AGS. It can hold up to twenty-one ready 105mm rounds and nine more in storage compartments in the front of the vehicle. It will be capable of loading a fresh round every five seconds, and should be quite reliable, based on the history of the naval gun mount and initial tests. Because of the autoloader, the crew of the XM8 has been reduced to just three, with the driver in the front hull, and the gunner and

commander on the right side of the compact turret. There is a thermal sight for the commander and gunner, as well as a 1553 data bus linking all of the vehicle systems. And plans are being made to possibly add an IVIS-style inter-vehicular network system. Powered by a 550-hp diesel engine with a hydromechanical automatic transmission, the XM8 will be capable of speeds up to 45 mph/74 kph, depending on the terrain and the installed armor package.

The armor is the big innovation on the AGS. In addition to the aluminum hull and turret, there is a layer of silicon carbide (the stuff that industrial drill bits are made of) tiles embedded in resin sheets bolted onto the hull to provide a level of protection similar to that of the -A2 version of the Bradley (called Level I protection). Additional overlay armor can easily be attached to the hull and turret to tailor the armor protection to the mission requirements and anticipated threat levels. Some of the add-on armor is composed of silicon carbide tiles, while other add-on plates are made of titanium or composites. When the Level III package is added, the XM8 is probably more survivable than the late models of the M60 Patton, which is still a formidable tank.

The reason for all of the armor options is that above everything else, the AGS is designed for deployability. Right now, it is just not practical to airmail an armored unit to someplace like Saudi Arabia without crippling the air transport system of the U.S. military. For example, when the 24th Mechanized Infantry went through the berms into Iraq, it had over 6,000 vehicles ranging from M1A1s to HMMWVs! With a C-5 Galaxy (our largest tranport plane) limited to carrying only a single M1 tank, and with only 150 of these priceless birds in service, it might take weeks to deliver a single armored division to the Middle East. But the XM8 is significantly smaller and lighter than an Abrams, and a C-5 can carry up to five with Level III armor installed, ready to roll off. The same C-5 can carry five XM8s with Level II armor and drop them by parachute into a drop zone. Within half an hour of hitting the ground, their crews (without any other supporting personnel) can have them ready to fight. And since a broken vehicle is about as useful as a dead one, the AGS is designed to be supported by a bare minimum of support personnel. For example, the three-

Interior view of the prototype XM4 Command and Control Vehicle (C2V). Note the extensive rack space for computers and radios, and the four very comfortable crew seats. *FMC Corporation*

man crew can remove the entire power pack in less than ten minutes, and can resupply all of the ammo in less than twenty.

The AGS will be the cutting edge of the Army's light forces in the 21st century. In addition to the tank batallions of the 82nd Airborne and 101st Air Assault Divisions, the XM8 will equip the new 2nd Armored Cavalry Regiment (Light). This regiment will provide scouting and screening for XVIII Airborne Corps (composed of the 24th Mechanized, 82nd Airborne, and 101st Air Assault Divisions). Currently, there are plans to buy about 800 XM8s, with requirements for perhaps a thousand more for the light infantry divisions and the Reserve/National Guard force. In addition, Taiwan has expressed an interest in buying some XM8s, so stay tuned for news on this neat little vehicle. It will be a handful when it arrives!

The XM4 Mobile Command Post

The venerable M577 has been around the U.S. Army since the 1960s; and even with the new -A3 version being fielded as of the mid-90s, the simple truth is that the Army needs a new mobile command post. To accommodate this requirement, the Army has contracted with FMC to build a new Command and Control Vehicle (called C2V by FMC) that has been type-classified as the XM4 Mobile Command Post. Built on the same robust chassis used for the MLRS launch vehicle, the C2V, now undergoing tests with the Army, carries a roomy box-like shelter. It is powered by a 600-hp diesel engine. Equipped with ballistic armor capable of stopping light machine-gun fire, and an overpressure/filtration/air-conditioning system, the XM4 will provide field commanders from battalion to corps level with a vehicle capable of conducting operations even while on the move.

The heart of the system is a flexible set of racks that can hold a variety of radios, computers, and electronic boxes. For example: an armored battalion/armored cavalry squadron command post might have four of the SINCGARS radios as well as command and control terminals tied to the regimental/brigade, company/troop, fire-support, and air-defense radio nets. These radios will be connected to a 33-foot/10-meter-tall erectable mast antenna, as well as to UHF whip antennas that allow the XM4 to command even while on the move. This helps reduce the vulnerability of the C2V to enemy artillery or air strikes that might try to home in on the XM4's radio transmissions. As for the six-to-eight-man crew, two of them will function as drivers/mechanics, and the rest will be in the back running various systems. And while the accommodations will not exactly be plush, the C2V will be the most comfortable vehicle in the Army. Though such comfort may seem a bit unfair to the line soldier, consider that a commanding officer may only get a few hours sleep a night while being expected to function at full efficiency. This can mean life or death for the soldiers under his command. For this reason, the environmental-control system on the XM4 is the most robust of any vehicle yet produced for Army service. And even the seats are designed to reduce fatigue and help the command crew stay sharp. FMC says that the chairs were modeled on the seats in airport rental-car shuttle buses, which, if you travel a great deal as I do, you know are some of the most comfortable in the world.

The XM5 Electronic Fighting Vehicle

During the 1980s, the Army modernized its battlefield electronic warfare (EW) capabilities. A major innovation was the EH-60 Quick Fix EW helicopter, used successfully during Desert Storm. The EH-60 can rapidly get directional fixes and then jam a variety of different communications, radar, and other electronic systems. Now ground forces will gain a similar capability with the XM5 Electronic Fighting Vehicle (EFV). Utilizing the same basic chassis as the XM4 C2V, it has the same basic enclosure. The only major external differences are a taller erectable antenna mast (98.4 feet/30 meters in height) and a different power system. The interior is roughly similar to the XM4, with similar racks of gear and the same comfortable seats and appointments. What is different are the types and missions of the "black boxes." These allow the crew of six to eight to scan for enemy transmitters, determine lines of bearing to the targets, pass the enemy positions along the intelligence chain of command (for possible attack by aircraft, helicopters, or artillery), and then, if desired, jam the enemy systems. While the Army and FMC decline to comment on the range of jamming options and how many channels each vehicle might be able to monitor and jam at once, it is clear that the XM5 will be a new ball game for the electronic warriors of the U.S. Army. And with many systems being software-driven, the XM5 will be capable of rapid reconfiguration to meet the changing tactical situations it will inevitably encounter.

U.S. Army Artillery Systems

Ever since the introduction of rifled tube artillery and mortars in the late part of the 19th century, no single class of weapon has caused more casualties than artillery. For no other form of weaponry has the ability to place a greater weight of explosive onto a target. By the close of World War I, artillery shells had grown in weight and capability to a point where proper use of artillery was seen as *the* key to the success or defeat of any attack or defense. And with the advent of chemical weapons by the Germans, and later by the Allies, a new and more deadly payload had been added to the capabilities of artillery systems. Since that time, nuclear warheads, land mines, cluster munitions, and even laser-homing warheads have come to be fired by the guns—or tubes as they are called by the Army—of the artillery. And the addition of rocket artillery and mortar systems has made the artillery even more powerful.

Shortly before the start of the 1991 Persian Gulf War, many analysts looking at Iraq's military focused a great deal of attention—and dread—on the huge Iraqi stockpile of tube and rocket artillery, as well as the huge supply of ordnance to fire from them. Armed with a wide variety of systems, from the awesome South African G5 155mm howitzer (designed by the brilliant but mercenary Dr. Gerald Bull) to huge Soviet multiple rocket-launcher systems, the Iraqi artillery could deliver anything from chemical weapons to small minefields. So dangerous was the threat that CENTCOM planners in the "black hole" planning center in Riyadh expended almost half of their "battlefield preparation" air strikes on eliminating the thousands of guns and rocket launchers in Kuwait and southern Iraq. And General Norman Schwarzkopf considered the elimination of the Iraqi artillery so essential that he insisted that Allied fliers eliminate at least half of it before he would start the ground offensive. In retrospect, this was probably wise. So for several weeks before G-Day, fliers from all of the Allies and services pounded the Iraqi artillery into silence. The result was that Saddam's vaunted artillery was scarcely felt by Allied forces on G-Day and after.

Meanwhile, from the very beginning of the air offensive, Allied artillery was integrated with the aerial bombardment effort. Whether it was an artillery raid over the border by Marine 8" self-propelled howitzers, or attacks upon enemy command and control bunkers by Army Tactical Missile System (ATACMS) missiles, the Allied artillery was swift, accurate, and deadly. Every time an Iraqi artillery battery opened up, an Allied artillery spotting radar would track the flight of the shells back to their point of origin, and quickly order a battery of Multiple-Launch Rocket System (MLRS) rocket launchers to destroy it, usually within less than a minute of the Iraqi battery firing its first shells.

Ironically, at a time when artillery has become more deadly than ever, the U.S. Army is phasing out many of the heavy guns that have been its mainstays since the Second World War. Today, monsters like the 175mm and 8" self-propelled howitzers are being retired in favor of new variants of the MLRS and M109. By 1997, only these two systems will be found in the heavy divisions of the Army. Improved performance will have to be based upon better software, data links, and superior warheads. Let's have a look at them and see how that is being done.

The M270 Multiple-Launch Rocket System (MLRS)

The use of rockets as artillery dates back to roughly the year 1000 AD, when the Chinese used primitive black powder rockets to frighten the horses of their enemies. Eventually, the Chinese, Mongol, and Indian armies had specialized troops whose job was to tactically employ rockets as bombardment artillery, incendiary devices, and possibly as signaling devices for messages between friendly troops. Much later, the British Army adopted an improved version of the black-powder rocket, designed by Sir Walter Congreve, for use in the wars of the Napoleonic period. Such rockets were used to telling effect during the Battle of Bladensberg (just prior to the British burning of Washington, D.C., in 1814), and the bombardment of Fort McHenry near Baltimore ("the rockets' red glare"). By the time of World War II, improved propellants and explosive payloads had begun to make the rocket into a really effective piece of artillery. The Russians and Germans produced the Katyusha and Nebelwerfer rocket systems for use in the Red Army and Wehrmacht, and both systems were much feared by their adversaries.

The U.S. Army used rocket artillery on a limited basis during the war, but the bulk of the postwar U.S. effort in this area went into guided missiles and nuclear-armed rockets like the Lance and Honest John. Only in the 1970s did the U.S. Army officially perceive a need for rocket artillery. Several factors drove the establishment of the new program that would become Multiple Launch Rocket System (MLRS). One of these was the extensive use of rocket artillery in the armed forces of the Warsaw Pact and Soviet client states around the world. A more practical rationale for acquiring the system was the highly effective performance of Soviet-supplied rocket artillery used by the Egyptians and Syrians against the Israelis during the 1973 Arab-Israeli War. The team of U.S. Army officers sent to survey the battlefields after the war noted the effects of the rocket artillery, and reported their findings back home. After reviewing the reports, Army officials decided to initiate a new rocket artillery program for the Army, and this became MLRS.

The Iraqis called it "Steel Rain," and the name was well deserved. Throughout the 1991 Persian Gulf War, any time an Iraqi artillery battery opened fire, an Iraqi command center began transmitting, or a high-value Iraqi target was found within thirty kilometers of the Allied front lines, ordnance fired from an MLRS would arch into the Arabian sky within a matter of seconds, and rain death and destruction on the Iraqi unit. Such was the power of this new weapon that, early in the war, it was often reserved for the purpose of eliminating Iraqi artillery firing on Coalition troops. Counterbattery fire it is called, and it was the mission of the MLRS batteries deployed by the U.S. and Great Britain to stop the Iraqi guns from firing once they were identified. At this task, and at many others, the MLRS proved to be the most flexible artillery system in the Allied arsenal.

The development of the MLRS system was one of the most trouble-free and efficient Army programs of the entire post-Vietnam period. Requirements for the system were developed by TRADOC; and in 1976, a program office was established at the Army Missile Command at the Redstone Arsenal in Huntsville, Alabama. In 1977, the program office initiated a competition between Boeing and LTV Aerospace (now Loral Vought Systems) to be the prime contractor for the effort. Another major program milestone occurred in 1979, when the U.S. Army signed a Memorandum of Understanding (MOU) with the United Kingdom, Germany, and France to produce the MLRS system for export to those NATO nations. Later on, Italy joined the MOU. The Netherlands, Turkey, Greece, Bahrain, Korea, Israel, and Japan are also MLRS users. In 1980, the LTV team was awarded the contract for integration and production of the overall MLRS system, which is built in Camden, Arkansas. The first units were delivered to the U.S. Army in 1982, with the first MLRS units reaching service later that year.

The MLRS launch vehicle is based on the M993 tracked carrier produced by FMC Corporation of San Jose, California. The M993 uses technology and components (such as suspension and power train) from the Bradley program, and has very similar driving and handling characteristics to the Bradley. The engine, a 500-horsepower Cummins V-8 diesel, drives a four-speed (three forward, one reverse) automatic transmission. The power train provides enough power to move the MLRS across the battlefield at a top speed of over 40 mph/64 kph. One interesting feature of the suspension is that the hubs are equipped with special locks (called suspension lockouts), so that the vehicle is more stable during rocket/missile launches.

The MLRS has the same ability to cross terrain as the other U.S. Army vehicles of its generation, and is air-transportable by a number of U.S. airlift aircraft. Unlike the Bradley, the armored crew compartment (bullet-proof against small-arms fire and artillery/mortar fragments) is in a cab at the front of the vehicle. The crew cab has room for three soldiers: typically a driver, gunner, and section chief. All crew members are cross-trained, and two soldiers can operate the system easily.

MLRS showing the rocket launcher at maximum elevation. The twelve rocket tubes are clearly visible in this view. *FMC Corporation*

The cab is equipped with an overpressure system to provide filtered breathable air during firing. And the crew remains in the cab for all firing operations. The crew cab is also equipped with a set of armored louvers to protect the windshield from being scorched by the rocket exhaust, or harmed by debris kicked up by the rocket blast. Inside, in addition to the driver's-side controls, are a terminal to provide fire control and communications to the Battery Control System (BCS) artillery-control network and two of the SINCGARS radios for voice and data communications. BCS is tied into the TACFIRE control system, which works at the division/regiment through squadron/battalion level to provide a clearinghouse for artillery requests—called "fire missions"—from front-line units. The fire-control portion of the system takes positional data from an inertially aided navigational system to provide "instant" firing options for the MLRS crew. And it is this ability to instantly react to fire commands that makes the MLRS system so special, as we will examine later.

The payload for the MLRS system is carried at the rear of the vehicle—the M270 multiple-rocket launcher. The launcher is hydraulically actuated and aimed by the fire-control system. It is capable of loading and carrying a pair of pre-loaded rocket/missile pods. The basic weapon of the MLRS system is a family of unguided rockets carrying a variety of different payloads to a maximum range of about 20 miles/32 km. The pod of six rockets is designated M26 (each MLRS launcher carries two M26s). The rockets, which are known as M77s, are 12' 10.8"/393.2 cm long, 8.94"/227mm in diameter, and weigh 676.5 lb/307.5 kg each. The M77 rocket is capable of carrying a number of warheads. These include the following options:

- M77—The primary warhead, and the only one used during the 1991 Persian Gulf War, is composed of a package of 644 M77 dual-purpose submunitions. Each submunition weighs about .5 lb/.25 kg, and is about the size of a hand grenade, with both a fragmenting case and a shaped charge for use against vehicles. As the submunitions are dispersed out of the warhead, each one trails a streamer to stabilize it and make the submunition pattern more predictable. Overall, a single M26 rocket pod will spread its payload of submunitions over an area of fifteen to thirty acres (depending on overlap). Exposed troops will theoretically be killed anywhere within the impact zone, and a direct hit from a single submunition will usually destroy a truck or soft-skinned vehicle or disable a light armored vehicle.

- XR-M77—During the Gulf War, it became apparent that the Iraqi Army had tube artillery (specifically, the South African G-5) with a range of up to 25 miles/40 km. Because of this, the U.S. Army decided to modify the M77 rocket to extend the range to 28 miles/45 km. This is accomplished by decreasing the load of submunitions by about 20% to 518 (down from 644), and increasing the length of the rocket motor.

- AT-2—A dispenser for AT-2 parachute-retarded anti-tank mines is being developed for the MLRS rocket by a German industrial consortium. The warhead contains twenty-eight of the AT-2s, each of which can disable a heavy tank.

- SADARM—Under the Sense-And-Destroy ARMor (SADARM) program, the U.S. Army has developed (Aerojet Electrosystems Co. is the prime contractor) a special munition which can sense the presence of an armored vehicle or artillery piece while descending on three small parachutes. The sensor is a dual-mode seeker (infrared and millimeter-wave radar), which can sense the "hot spot" of a tank's engine compartment, as well as the vehicle's "center of radar mass," which is normally the turret structure. Each rocket carries six of the SADARM munitions (there is also a program to deploy a 155mm tube-artillery shell capable of deploying two smaller SADARM munitions), which are ejected over concentrations of vehicles. When the sensor detects an armored vehicle or artillery piece in its field of view, it fires a self-forging projectile down into the top of the target. The status of this program is in doubt at the moment due to budget cuts.

- TGW—Another "smart" warhead for the MLRS was the Terminally Guided Warhead (TGW) munition. The TGW was a large anti-armor munition which used a "smart" millimeter-wave radar seeker to search out tanks and other priority targets. Once a TGW munition recognized a valid target (it can discriminate between various types of targets such as tanks, IFVs, artillery, etc.), it maneuvered over, and dove into the top of the target, destroying it with a large shaped charge. Each rocket would carry three of the TGW munitions. The United States opted out of the

TGW program; France, Germany, and the United Kingdom are struggling to maintain it.

While there are many initiatives to develop and deploy improved warheads, only the basic M77 is in service. And it was this version that went to war in 1991. Overall, 201 (189 U.S., 12 British) MLRS vehicles fired a total of 9,660 M77 rockets during Desert Storm. These delivered some 6,221,040 of the deadly submunitions on Iraqi targets. The commander of the one British MLRS battery (twelve launchers) called it "the grid exterminator," because of its ability to completely devastate an entire grid square (a square kilometer on standard military maps).

Good as unguided rockets are, there are times that a field commander wants to hit targets farther away than the 20-mile/32-kilometer range of the current variants being deployed. Many of the things that might cause a commander concern—such as surface-to-air missile (SAM) sites, command posts, and logistics centers—are usually deployed in what is commonly known as the "rear echelons." Up until recently, the only options open to an Army commander to strike at such targets was to call the Air Force or Navy for an air strike, or risk a high-value asset like an AH-64 Apache attack helicopter or a special-operations team to destroy them. In 1991, though, the Army deployed to the Persian Gulf 105 units of a new surface-to-surface weapon (SSM) known as the Army Tactical Missile System (ATACMS). ATACMS is a stubby little SSM about 12 feet/4 meters in length, and 2 feet/.6 meters in diameter, with a range of between 60 and 90 miles/100 to 150 kilometers. It carries a payload of 950 M74 cluster bomblets, each of which roughly approximates a hand grenade in its effects. There are plans for other types of warheads for use against special targets.

Built and assembled at several facilities in Texas by Loral Vought Systems, each ATACMS is packed into an MLRS pod (which normally holds six M77 rockets), and can be loaded and fired by any MLRS launcher that has been modified with the appropriate fire-control and software upgrades. What makes ATACMS different is more than a simple extension in range. ATACMS has a highly accurate guidance system, which allows it to place its cloud of M74 bomblets exactly on target. And while the M74s do not quite pack the punch against armor that the M77 submunitions do, they are optimized for effects against "soft" targets. This means that things like radar sites, command-and-control vans, trucks, and fuel dumps are quite vulnerable to ATACMS missile strikes. This is not to say that the Army plans to dump ATACMS missiles all over the battlefield. If their use in Desert Storm is any indication, the U.S. Army tends to treat them like "silver bullets." In fact, only thirty of the 105 ATACMS missiles that were shipped to the Persian Gulf were fired by the eighteen M270 launchers that had been modified to fire the new missiles. The first shot fired in combat by VII Corps was a single ATACMS fired at an Iraqi SA-2 Guideline SAM battery. The battery was destroyed. The key point of this mission is that the Air Force called for the strike on the first day of the air war, to make it safe to fly their missions over the target zone!

So just how does one make use of all the technology and firepower embodied by the MLRS system? Well, consider the following example. An American force is massing for an attack on an opposing enemy ground force. The other side is not as stupid as Saddam Hussein, and they begin to detect signs of the coming attack. In response, the enemy commander plans a night artillery attack with heavy-tube artillery on the American force as it is concentrating, and thus when it is most vulnerable. The first sign is the flash of the enemy guns on the horizon. The next signal is more useful to the Americans. Somewhere along the American line sits an observation post with a Q-37 Firefinder artillery-spotting radar. The radar, which is designed to track the incoming shells back to their positions of origin, rapidly plots the positions of the enemy batteries on a terminal connected to the TACFIRE artillery-control network. So quickly does the Q-37 crew do their job, that all of the enemy batteries are probably plotted before the first shells have hit the ground. Up at the headquarters of the American unit, the TACFIRE computer is allocating fire missions for artillery units that have been programmed to stand by for such assignments. In fact, all of the fire missions for these units will probably be delivered electronically before the second rounds are fired by the enemy batteries.

Things really begin to happen quickly now. Each battery commander with a counter-battery mission will execute it as quickly as is possible. The quicker the enemy batteries are silenced, the fewer Americans will wind up in the hospital or in body bags. So vital is this mission that the current U.S. Army standard for responding to a counter-battery fire mission is about one minute flat. If the MLRS battery commander is already stopped and emplaced, then all they need to do is set the position of the enemy battery into

Interior of MLRS crew cabin. The gunner is entering targeting information on a keyboard. The large guarded toggles to his right are safety and arming switches. Both crewmen are wearing standard CVC (Combat Vehicle Crew) helmets with built-in microphones. *FMC CORPORATION*

the fire-control system and fire. But one of the best things about the MLRS system is its ability to move around in support of mobile units. So let's assume that the unit we are looking at has been on a road march to the front. Each battery has a total of nine launchers (in three platoons of three), as well as eighteen M985 artillery ammunition carriers with trailers. These are variants of the M977 HEMTT truck, and each one tows a trailer equipped to carry the rocket pods for the MLRS system. Each M985 truck can carry eight pods of rockets, and each trailer carries eight more. In this case, we'll assume that two of the platoons are loaded with M26 rocket pods (twelve rockets per launcher), and the other section has been loaded with ATACMS missiles (two per launcher). The battery commander decides to use only the six vehicles loaded with rockets, and hold the three launchers with their load of ATACMS missiles in reserve.

As soon as the fire mission is received, the battery commander orders the ammunition carriers to get away to a safe distance (to protect them from the rocket blast and any enemy counterfire), and the battery sections to deploy. Because each vehicle has its own POS/NAV system being fed updates from a NAVSTAR GPS receiver, each vehicle commander, and the onboard fire control system, can determine the position of each launcher with an accuracy of plus or minus 16 feet/5 meters. This also means that the launchers don't have to be sited near each other to get an accurate dispersion pattern from the warheads of the M77 rockets. This data is automatically fed into the fire control system, as well as the position of the target. The gunner's only job at this point is to monitor the system and tell it just how many rockets it is to fire (any number from one to twelve rockets per vehicle is possible), and await the fire command from the battery commander. The driver is somewhat busier, setting up the vehicle to fire. He locks the suspension to hold the vehicle steady, lowers the armored louvers over the windshield to protect them from the coming blast, and turns on the cab overpressure/filtration system to protect the crew from rocket exhaust fumes. While this may sound like a lot, it all happens in just a matter of seconds, and the MLRS is ready to fire.

When the six firing vehicles indicate that they are ready, the battery commander orders them to open fire, probably within no more than a minute after having received the fire mission from the TACFIRE system. The armored box containing the rocket pods at the rear of each launcher swivels and tilts to the proper angle and bearing (as set by the fire-control computer), and the first rocket is fired, followed at intervals of about five seconds by the others in a "ripple." As the motor fires, and the rocket moves down the launch tube, it rides on a set of helical rails that impart a slow spin to help stabilize the rocket when it is ejected free. After it emerges from its launch tube, a small pyrotechnic charge helps to deploy and lock into place the four curved fins at the rear of the rocket. The curvature of the fins makes the rocket continue to spin. Also, the launcher automatically re-aims itself after launching each rocket to make up for any "jiggle" that might be caused by the rocket back-blast. (Crew who have fired the rockets tell us that the sound inside the vehicles is

ABOVE LEFT: MLRS firing a rocket at the White Sands Missile Range in New Mexico.
OFFICIAL U.S. ARMY PHOTO

ABOVE RIGHT: MLRS Rocket Firing Sequence #2.
OFFICIAL U.S. ARMY PHOTO

LEFT: MLRS Rocket Firing Sequence #3.
OFFICIAL U.S. ARMY PHOTO

like the inside of a thundercloud. And to nearby observers, the firing of MLRS rockets seems to sound like ripping glass. Whether you're inside or outside, the sight is impressive as the rockets fly out, particularly at night.) After the HTPB propellant rocket motor burns out a few seconds later, the rocket follows a ballistic course the rest of the way to the target. Once the rocket is over the target area, the electronic fuse detonates the dispersion, or core charge, at the center of the warhead. Then the M77 submunitions, which are packed in polyurethane foam, are dispersed and fall in a cloud towards the target. Since the six MLRS vehicles deliver a total of some 46,368 M77 munitions onto the enemy artillery site, the results are horrific: In all likelihood, every gun and artillery tractor will be struck by one of the submunitions, and either damaged or destroyed. Vehicles containing ammunition will have their loads detonated by contact with the M77s. And it goes without saying that the artillery crews are probably not going to survive. In fact, all that will probably be left at the site will be scrap iron and shredded flesh. This all happens before the guns have finished firing their third or fourth shell. All along the front, enemy artillery units simply "dissolve," without so much as a whimper from their commanders. In a little while, the enemy high command will call, but nobody will be home to answer.

But this is hardly the end of the story, for the battery commander now has the problem of making sure that what has just been done to the enemy battery does not happen to *his* battery. Ordering the battery into motion, he selects a grid coordinate where the battery can reform and reload the now-empty launchers. (This is what the U.S. Army means when it describes MLRS as a "shoot-and-scoot" system.) Meanwhile, as the battery heads for the reload point, other events are under way. Several American radio-direction-finding (RDF) units have been tracing the communications signals from a number of rear-area (within 60 to 90 miles/100 to 150 kilometers of the front line) enemy command posts (CPs); and the American force commander has decided to deal with them now. Once the positions of the enemy CPs have been determined, each is assigned a fire mission from an ATACMS missile launcher. Once again, the TACFIRE system sends a string of instructions down to the battery to the onboard BCS terminals. The battery commander then likely just peels off the three ATACMS launchers to the side of the road, where they fire and then move along to the reload site on their own.

The setup procedure for the crew to fire ATACMS is almost identical to that of the M26 rockets. Only the end result is slightly different. Once again the launcher is locked down and sealed, the target positions are fed automatically into the fire-control system, and the launch signals given. Since ATACMS are valuable and their quantity limited, only one missile will be fired per target. As each missile clears the launch pod, the guidance fins deploy, and the missile heads off to its target looking for all the world like a cartoon weapon (ACME: a name you can trust!). The ATACMS is so stubby and bloated that a painted shark mouth (like the Flying Tigers used to paint on their P-40s) would not be out of place on it. Meanwhile, using an inertial guidance system based on a special ring laser gyro, the ATACMS maneuvers so that it will arrive directly over its target. It takes several minutes to fly the 60 to 90 miles/100 to 150 kilometers. Once there, a small core charge detonates inside the package of cluster bomblets within the warhead. The bomblets then disperse over the target area, each with the destructive power of a hand grenade or small mortar round.

For a lightly protected target like a headquarters unit, this will prove extremely destructive, and will probably destroy everything except armored vehicles. The most likely sign of success for the ATACMS strike is the enemy headquarters radio circuits going silent, never to flash their electronic signals again. As for the MLRS battery, it will by this time have moved to its reload position, and be re-arming the launcher vehicles. This is accomplished by having the M985 winch a pair of loaded rocket/missile pods to the ground behind each launcher. Once this is done, the launchers' own built-in loading gear winches each pod up into the M270 launcher, and the battery is ready to go back into action.

As you can see, the MLRS system has features that make it one of the most flexible and powerful weapons in the Army inventory. By late 1993, some six hundred launchers had been delivered to the U.S. Army, with more in the hands of the MLRS consortium partners. The Army is currently planning to field almost 1,300 launcher vehicles, and maintain a large inventory of MLRS-compatible rockets and missiles, though budget cuts may force this number much lower than planned. Nevertheless, it is the truth of the moment that

MLRS is the finest artillery system currently in use by the U.S. Army today. And until the M109A6 Paladin 155mm self-propelled gun comes on-line later in the decade, it will remain the longest-ranged, fastest, and most accurate artillery system in the Army.

The M109A6 Paladin 155mm Self-Propelled Howitzer

Under General Sullivan's plan to modernize the Army by 1997, even the fairly mundane-looking and comparatively old-fashioned tube artillery is getting a digital makeover. In fact, all the qualities that make the MLRS system so useful—speed, responsiveness, positive navigational accuracy, and inter-vehicle connectivity—are currently being planned for the next version of the U.S. Army's self-propelled howitzer (SPH), the M109. The reasons for this upgrade are simply that tube artillery can (still) deliver a greater weight of accurate firepower on a target than any other weapons system. It is efficiency, simple and pure, that drives U.S. Army SPH development, and that efficiency is the reason for the upgrade program known as the M109A6 Paladin. It is an efficiency based upon the historical utility of self-propelled artillery. This particular development in the field of ballistic weapons has an important past, and a vital future. Let's take a look.

Big guns are heavy. When mobile artillery was first conceived in the 15th century, artillery pieces had to be hauled by large teams of oxen. While an ox is slower, meaner, and dumber than a horse, it has more strength and endurance for the long haul. But by the 17th century, guns and carriages could be made strong and light enough for a pair of horses to pull them along, though the gun crews still had to slog along on foot. In the late 18th century, some armies began to mount the gunners on their own horses, and they were accompanied by light two-wheeled carts called "caissons" to carry gunpowder and projectiles. This "horse artillery" could move at cavalry speeds, and was drilled to get the guns in and out of action ("limbered" or "unlimbered") extremely quickly. By World War I, trucks or tractors began to replace the horse, and rubber tires replaced the iron-rimmed wagon wheel. But at the start of the Second World War, most armies still had some horse-drawn artillery. With a motorized carriage, a force could move guns at highway speeds, at least until the highways turned to mud. Then the guns, like all the other things dependent upon wheeled transport, bogged down, a victim of "General Mud."

Clearly, a superior form of transport for artillery was needed. The solution was tracked, self-propelled artillery. The World War II German Wehrmacht was the first to systematically introduce self-propelled artillery by pulling the turrets off obsolete tank chassis and rigging improvised mountings for field guns. The gun, the crew, and some ammunition could now move cross-country and keep up with the Germans' advancing Panzer units. Unfortunately, these early open-topped systems provided no overhead protection for the crew against shell fragments—or for that matter, against the rain! The Allied forces also developed self-propelled artillery pieces during World War II, notably the M3 "Priest," with a small 105mm pack howitzer in the mount. By the 1950s, the U.S. and British Armies had begun to introduce self-propelled 105mm guns with fully enclosed turrets. These vehicles looked like tanks, though the bulky turrets had a limited traverse (they could not rotate 360 degrees, like a tank turret). To stabilize the chassis against the recoil of the gun, some models required the use of a "spade"— a broad spiked blade on a hinged mounting at the rear of the chassis. This was lowered before firing to dig into the ground and thus stabilize the vehicle against the gun's recoil.

The design of a self-propelled gun, like that of any other armored fighting vehicle, involves trade-offs between mobility, protection, weight, speed, ammunition capacity, and mechanical complexity. National style and doctrine also play a role. For example: To save on scarce manpower, the Swedish Army designed their 155mm self-propelled gun with a sophisticated mechanical autoloader; Soviet-designed self-propelled guns are designed to use flat-trajectory, direct fire to simplify fire control and provide a secondary anti-tank capability.

The origins of the American M109 program date back to the 1960s, when the Army decided that their next generation of SPH would put the gun, crew, and ammunition under armor to protect against counter-battery and small-arms fire. The initial results of this program were the M108 (with a 105mm howitzer) and the M109 (with a 155mm howitzer). These were guns on a tracked chassis, with an armored box or cab built over the top of the chassis. The armored cab had room for the crew to load and fire the howitzer, as well as stowage for ammunition.

These relatively simple vehicles were adequate for service in the Vietnam War and the early days of the Cold War, but new operational needs emerged in the 1980s. One was the need to accurately survey and locate the site used as a firing position, without leaving the protection of the vehicle (the more accurately you know where you are, the more accurately you can aim at your target). In addition, the time required to set up a battery and establish secure commu-

The M109A6 Paladin Self-Propelled Howitzer (SPH).
JACK RYAN ENTERPRISES, LTD., BY LAURA ALPHER

nications determined the unit's ability to respond to an immediate request for fire. As late as the Persian Gulf War of 1991, the U.S. Army standard for setting up an M109A2 battery (eight M109s along with their ammunition carriers) was around eleven minutes. Compared to the responsiveness and performance of the MLRS batteries, which measured their response time in seconds, tube artillery in the Gulf War seemed sluggish. But even before the forces of the United States began to deploy to the Gulf in 1990, the Army was already well into the development of a new generation of M109, the M109A6 Paladin.

The idea behind the development of the Paladin is similar to the one that led to the M1A2 Abrams. Take an existing vehicle chassis (such as an early-model M109A2 or M109A3), strip it down to the bare bones, rework all of the automotive and suspension components (essentially "zeroing" the vehicle from a "wear and tear" standpoint), replace all of the old vehicle electronics (called "vetronics" in the jargon of the industry) with newer, more reliable digital systems, tie them all together with an onboard data bus, and add the necessary digital communications, navigation systems, and a fire-control computer to provide rapid response to requests for artillery fire. In this way, the Army can take a gun that is essentially obsolete and make it into the world's best tube-artillery system in one major upgrade.

As you walk up to a Paladin, there are few external details to tell you it is different from any earlier M109 model. Only the larger turret bustle and additional radio antennas mark it as something new. Inside is another thing.

Normally, the crew enters the vehicle from the rear by ducking under the bustle overhang and climbing through the main hatch. This normally is done on hands and knees. But once you're inside, as you stand up, you get the feeling you could play handball in there. There is a lot of room. Unlike inside the tight confines of an Abrams or Bradley, even someone over six feet can stand and move around comfortably.

The interior layout of the Paladin is much like that of the Abrams, with the driver up forward, the vehicle commander and gunner positions on the right side of the gun turret, and the loader on the left. Lining all of the interior surfaces is a Kevlar spall liner designed to reduce the danger of spall fragments to the crew. To the rear of the turret is the storage area for ammunition and propellant charges. The Paladin has room in the rear bustle to store a total of thirty-seven NATO Standard 155mm rounds, a pair of Copperhead laser-guided projectiles, and the necessary propellant charges to send them on their way. Some of the different ammunition types include:

- High Explosive (HE)—A steel-jacketed case filled with PBX-series explosive. Impact-fused, it detonates on contact with the ground, with blast and fragmentation effects.
- High-Explosive, Variable-Time (HE-VT)—An HE shell fitted with a radar proximity fuse. The fuse can be set to explode at a particular height over the ground, showering the target with fragments when it detonates.

- White Phosphorus (WP)—Called "Willie Pete" by the troops, this round has a case filled with chemical white phosphorus and a small bursting charge. When the shell hits the ground and the burst charge detonates, the white phosphorus spontaneously ignites in the presence of oxygen (as in air or water). In addition to incendiary effects, it generates a dense cloud of white smoke that can serve as a target marker for other weapons. The downside of WP is that it is dangerous to manufacture and handle. And the terrible burns it inflicts have led some people to regard it as a particularly atrocious and inhuman weapon. In spite of this, virtually every nation with artillery uses WP rounds.

- Field Artillery Containerized Anti-Tank Mine (FASCAM)—This is an artillery shell containing a number of anti-tank mines capable of blowing the track off an armored vehicle or destroying a wheeled vehicle like a truck. FASCAM rounds can be lobbed into an area like a crossroads or a pass, where they can block the movement of an enemy force until they are cleared.

- Smoke—Frequently, smoke rounds (shells with pyrotechnic smoke generators in them) are carried as a means of marking a position or masking movement from an enemy. Smoke rounds come in a variety of colors, such as red, blue, purple, and orange.

- Jabberwocky—This projectile, named after Lewis Carroll's poem, is a shell containing a powerful little broadband-radio noise-jammer designed to disrupt enemy communications. When the Jabberwocky is over its target area, it deploys a parachute to retard its fall and soften its landing. After landing, it deploys an antenna and starts jamming. A small thermal battery built into the shell case provides power for a few hours.

- SADARM—Like the M77 MLRS rocket, there is in development a 155mm artillery shell capable of carrying and dispersing several SADARM "skeets" to destroy enemy tanks and vehicles. This program appears to have become a victim of budget cuts, and a replacement called BAT (with an acoustic sensor) is slated to take its place.

- Copperhead—This is the crown jewel of American artillery. Designed for use against tanks and point targets, the Copperhead is the artillery equivalent of a laser-guided bomb. When it is fired, small guidance fins pop out from the body of the shell, and a laser seeker in the nose starts searching the ground for a spot of laser light pulsing with a specific code. This is the signal from a laser designator which is marking—or "painting"—the desired target. The laser designator, with a range of several miles/kilometers, can be a hand-held unit operated by a forward observer, or it can be mounted on a combat vehicle or helicopter. When the seeker detects the laser spot, it homes in, adjusting its course with the guidance fins.

The high-explosive warhead detonates on contact, and is sufficient to destroy a tank. The only problem with the Copperhead is that the supply is limited. Because of cost, the production line was shut down before completing the planned production run, and Copperheads have to be used sparingly. Nevertheless, this may be the most accurate and deadly point-target artillery weapon in the world today.

This is not a complete list of everything that can be stuffed in a 155mm shell. It is only a general introduction to the capabilities of U.S. Army artillery. Meanwhile, some of you may be wondering about the exclusion of nuclear and chemical weapons from the list of artillery munitions. The simple truth is that such weapons are a thing of the past. At one time, the Army had a 2-kiloton-yield nuclear artillery shell called the W82, but in 1991 President Bush ordered the removal of all nuclear weapons from deployed U.S. forces. In addition, U.S. forces also deployed a small but powerful array of chemical weapons. Luckily, the program was terminated as a result of a series of international treaties; and the stockpiles are gradually being destroyed at incinerators in Kentucky, Utah, and Johnston Atoll in Micronesia.

Structurally, the Paladin is mainly built of aluminum (like the Bradley and the M113), though significant portions are still composed of high-quality steel. Weighing in at 63,300 lb/28,181 kg, it is powered by a 440-hp turbocharged Detroit Diesel V-8 engine driving a six-speed (four forward, two reverse) transmission. This provides for a top speed of 35 mph/57.4 kph and a cruising range of 214 miles/351 kilometers. The business end of the Paladin is an M284 155mm, 39caliber cannon mounted on a M182 gun mount. It is capable of firing conventional ammunition out to a range of 14 miles/23.1 kilometers, and rocket-assisted rounds out to 18.3 miles/30 kilometers.

What makes the Paladin so different are all of the systems that have been improved or replaced. The major upgrades include:

- Suspension System—With the weight of the Paladin growing to almost 64,000 lb/29,091 kg, the suspension of the original M109 would have a difficult time maintaining stability during cross-country dashes from firing position to firing position. To make up for this, the M109A6 has been designed with longer torsion bars and hydropneumatic bump stops to smooth out the ride and provide greater support for the increased load of the new vehicle.

- Automatic Fire-Control System (AFCS)—The AFCS provides an integrated solution to the problem of determining vehicle position, receiving fire missions, and automatically pointing the gun at a target. It has an inertial navigation system called Modular Azimuth Positioning System (MAPS). In addition, the system can receive targeting data over the onboard radios directly into the fire-control system. This means that all the crew needs to do to put rounds on target is to fit the proper fuse onto the nose of the shell, load the shell and propellant bag into the weapon, and pull the firing lanyard. And while MAPS currently lacks a NAVSTAR GPS receiver, it is likely that this will be installed in the very near future.

- Voice/Digital Communications Systems—Where previous versions of the M109 were limited to receiving their fire missions via a landline from the battery operations center, the Paladin has a pair of AN/VRC-89 SINCGARS radios to provide both secure voice and digital communications, as well as a data link to the TACFIRE/BCS system.

- Automatic Remote Travel Lock—Because of the length of the M109 gun barrel, it is necessary to lock it down during movement to relieve stress that might warp or distort the tube, affecting gun accuracy and safety. Previous versions of the M109 required a crew member to exit the vehicle and manually lock the barrel into its travel lock. On the Paladin this is done automatically in less that fifteen seconds.

- Driver's Night Vision Device—Previously, the M109 had difficulty making night marches and participating in night artillery raids, due to the lack of any onboard night-vision system. This shortcoming has been rectified in the Paladin through the addition of a an AN/VVS-2(V4) night-vision system. This is a "clip on" viewer which can be installed rapidly (within fifteen seconds) without tools on the driver's periscope. It is a relatively new system that uses light/image intensification to provide a clear outside view, even on moonless nights.

- Microclimate Cooling System (MCS)—Under the best of conditions, the firing of a 155mm howitzer is hot, dirty, and dusty work. When the crew is wearing MOPP-IV chemical-protection suits, it can be downright debilitating. To overcome the heat and fumes, the Microclimate Cooling System (MCS) provides filtered and cooled air to make the wearers more comfortable. Each crew member is attached to the MCS by a hose-fed face mask. Even

Field Artillery Ammunition Supply Vehicle showing the fold-out power-operated conveyor system that can be used to transfer ammunition to an M109. *BMY COMBAT SYSTEMS*

when the crew is not wearing MOPP-IV suits, they wear the masks to avoid inhaling propellant fumes from the gun.

- Electrical Power System—All of these electronic "goodies" require a lot of power. To accommodate the increased power requirements of the M109A6, a 650-amp alternator has been installed to support the increased electrical load.

What all of this means is that the M109A6 is able to do with a cannon almost everything that the MLRS system does with rockets and missiles. And it can do it with similar mobility and speed. It is designed to be part of the 21st century digital battlefield. And unlike earlier versions, which had to stay behind the front to survive, the Paladin can move with the most forward elements of an armored force. At a moment's notice, it can stop in place and have the first round "on the way" in less than sixty seconds. The ability to "shoot and scoot" increases the system's survivability against counter-battery fire.

There is a special vehicle designed to keep the Paladin full of ammunition and ready to shoot. Called the M992A1 Field Artillery Ammunition Supply Vehicle (FAASV), it carries ammunition, propellant, and fuses to Paladin units at the front lines. FAASV is based on an M109 chassis, with an armored box that holds up to 90 shells, 3 Copperhead laser-guided rounds, 99 propellant charges, and 104 projectile fuses. The M992A1 can move with the Paladin, operate in the same area, and even resupply it while under fire.

The FAASV has a conveyer system that can pass ammunition and propellant charges between it and an M109A6 without requiring the crews to exit the vehicles. This makes a big difference should a Paladin battery come under enemy fire (though under safer conditions, crews usually prefer to move the ammunition and propellant charges between vehicles by hand). In addition to these capabilities, the FAASV can be used as a prime mover for virtually any of the towed howitzers in the Army inventory should it be necessary. It even has the muscle to tow a disabled Paladin!

Each of the 824 Paladins that the Army plans to buy will have its own FAASV assigned to support it in the field. Older self-propelled guns will retain ammunition carriers such as the M548, which is based on the M113 chassis.

So just what does all of this mean to the crews who will operate the Paladins and the commanders who will depend on it for responsive, accurate artillery fire? Well, consider the following example. One of the most popular uses for mobile artillery during Desert Storm was the armored artillery raid. These were hit-and-run raids into Iraq and Kuwait by self-propelled artillery to attack targets that required something more than a few bombs to destroy, but were beyond the range of the 16"/406mm guns of the battleships *Wisconsin* and *Missouri*. So valuable were these raids that they appeared on the daily air tasking order issued from the "black hole" headquarters in Riyadh which controlled all bombardment of Iraq and occupied Kuwait.

So let us assume that we have a battalion of Paladins that has been assigned to move across the front lines to conduct such a raid on an enemy fuel depot some 30 miles/50 kilometers behind the front. Their job will be to rapidly move about 18 miles/30 kilometers into enemy territory, rapidly set up, and fire a dozen or so rounds from each vehicle. They then will have to quickly move to another position to repeat the procedure, and then run for home across their own lines.

An artillery battalion never sets off on such an adventure without help and a lot of planning; and so this mission is laid out to the last detail. The battalion itself is composed of three artillery batteries, each of which is composed of eight M109A6 Paladins (organized into two platoons of four vehicles each), eight M992A1 FAASVs (four per platoon), and a pair of modified M577s called Platoon Control Vehicles (PCVs—one per platoon). The PCVs are designed to take fire missions from the TACFIRE system and rapidly distribute them among the M109A6s in the platoon. If necessary, however, Paladins can accept and process fire requests directly from a forward or aerial observer. According to the Army's usual practice, the guns are escorted, in this case by an armored cavalry troop (nine M1A2s, thirteen M3 Bradley scout vehicles, an M981 FIST-V, and a pair of M125 106mm mortar carriers) and a few OH-58D Kiowa Warrior scout helicopters to guard the task force's flanks. Not so large a force that it attracts a lot of enemy attention, but large enough to destroy the enemy supply base or blunt the nose of an enemy offensive probe.

The artillery task force sets out after dark, with the scouts of the armored cavalry troop leading the way, and with the Paladins and PCVs behind them. Seeking a seam in the line where enemy coverage is thin, the force moves through as quickly as possible to minimize their exposure to enemy observation. Thanks to the IVIS system on the cavalry vehicles and the interface (a sort of digital "hook") to the TACFIRE system, all of the maneuver orders are issued without any need to communicate via the voice radios, and thus attract enemy attention. Also, with every vehicle having a precision navigation system, there is no need for them to travel together. In fact, the task force could break into small platoon-sized groups and move separately to the first assembly area. Meanwhile, the FAASVs and a few of the tanks and scout vehicles lag behind to set up a resupply area, ready to move wherever the artillery task force needs them.

As is normal in such operations, several small two-man teams of Army Rangers from the Special Forces Group at Fort Benning, Georgia, are placed ahead as scouts and forward observers. These teams observe any enemy movement in the target area, and they will provide corrections for the artillery strikes. The move to the first fire position is quick, taking less than an hour from the time of crossing the front line. As soon as Paladins are in their planned firing positions, each vehicle sends a data-link signal up the network to indicate it is ready to fire. What happens next will be fast. Very fast.

In the task force commander's vehicle, a message is tapped out on a data terminal (called a "Data Message Device" or DMD) calling for a fire mission on the fuel depot. Each platoon is assigned a different part of the facility. At the end of the message is a final piece of data, the "time on target" (TOT) for the artillery barrage. To maximize the effect of surprise, the commander tries to synchronize the firing of every gun, so that the first rounds all impact simultaneously. Once the fire mission is on the way over the network, and the TOT is being counted down, the breeches of the howitzers are opened, projectiles (conventional high explosive for this job) and powder bags are rammed in, the breeches closed, and the final checks made. The Paladin commanders and gunners punch in the target coordinates, and the MAPS system continues to update gun location and direction, calculating just where to point the gun in elevation and azimuth. The crews, breathing through their MCS masks, are already preparing the second shot when the fire-control system fires the first round. In less than a second (depending upon their positions relative to the target area), all twenty-four Paladins have fired their first round, and the crews are loading their second. Normally, the Army standard is for a crew to be able to load and fire one round per minute until the ammunition runs out. But there is also a rapid-fire drill, with the crew doing a frantic but precisely orchestrated dance inside the turret of the M109A6, firing around a dozen rounds in just over three minutes. This is what happens now. In a little over 180 seconds, some 288 high-explosive shells are exploding on the enemy fuel depot some 12 miles/20 kilometers away. Meanwhile, the team of Rangers, in a concealed overwatch position, are tapping out battle-damage assessments on a small telecommunications terminal, to inform the task force commander if the target has been destroyed, or if it has to be hit again. As the fuel dump goes up in flames, it is time to go home.

"Shoot and scoot" is the name of the game here, and that is exactly what happens now. As soon as the last of the rounds have left the gun tubes, the Paladins, PCVs, and their cavalry escorts are running to the secondary fire position that has been laid out for them. Only about a ten-to-twelve minute dash from the primary position, the Paladins stop here and see if they need to fire more upon the depot, or to take action to insure their own escape. In this case, the decision is for a quick 12-round fire mission of FASCAM mines upon the probable enemy approaches to their escape route home and to the depot (to keep them from fighting the fires). This done, everyone heads by planned routes back to friendly lines. Along the way, if one of the OH-58Ds Kiowa Warrior scout helicopters sights anything pursuing the retreating Paladins, the Paladins themselves can be their own best help, with a quick fire mission of Copperheads. All the gunner/observer on the chopper need do is lay the laser designator on the targets, punch in a designator code, feed the request for fire mission onto the network and, within sixty seconds, a Copperhead will be arching over the battlefield for a direct hit on the intruder.

Since there was no need for extra ammunition, the FAASVs with their escorts are now headed back for a rendezvous with the task force Paladins back behind friendly lines. Each M109A6 is given a grid coordinate to meet up with its assigned FAASV, and then the process of resupply and clean-up begins. For the gunners and loaders, this means getting out the cleaning rods and Break-Free cleaner to clear the powder residue from the gun barrels and breeches, and restocking the ammo and propellant racks. For the driver, it is maintenance of the tracks and checking the vehicle fluids. For the commander, mission and ordnance reports, and starting to get ready for the next day's fire missions.

Such is the daily life for what has become known as the "king of battles," artillery. And right now, the M109A6 rules the kingdom.

The Future

We have spent a lot of words looking at the importance of artillery on the modern battlefield. If this has proven anything, it is that the long reach of artillery can "vaporize" entire enemy units at the flick of a switch. And while other types of fire such as that from helicopters and tanks can reach out and kill a single vehicle with deadly accuracy, artillery can kill many beyond visual range in a single strike. Though this might seem to be enough, there are already initiatives afoot within the Army to make cannons even more deadly. The first of these is a new artillery-control system. While the TACFIRE net has been a capable control system, it suffers from several shortcomings. One is a shortage of interface "hooks" with other maneuver and vehicle-control systems. Another is the dependence of TACFIRE upon large, centrally located computer processors at the battalion artillery command post. The destruction or failure of the TACFIRE processor means the loss of the whole system served by that processor. To get around this, the U.S. Army is starting to deploy a new artillery-control system called the Advanced Field Artillery Tactical Data System (AFATDS). AFATDS is designed to overcome the TACFIRE's limitations, while adding new capabilities that had not even been imagined when TACFIRE was designed. For starters, AFATDS is composed of numerous, smaller computer processors that work in what is called a "distributed" architecture. This means that a computer processor can be placed anywhere on the AFATDS network and contribute its power to the job of running the network. In addition, the AFATDS consoles and user interfaces have been completely redesigned, so that they are easier to use and can be rapidly reprogrammed to accommodate new tactics and artillery systems. AFATDS has also been designed to work with and "talk" with a wider variety of U.S. Army systems than TACFIRE, so that more people in a given unit can call for and receive artillery support when and where they need it. AFATDS is a long-overdue capability for those who manage the big guns.

As for the guns themselves, there are several possibilities. The U.S. Army has been doing research and development on an Advanced Field Artillery System (AFAS). This is a new kind of howitzer that uses a liquid propellant (LP) instead of the bagged solid propellants used today. LPs have the advantage of being both more efficient and more capable of being controlled than pre-measured bags of solid propellant. The downside is that storage and management of the LP are more difficult, and this has required some very advanced chemical engineering just to make it work. Some in the Army argue that it is a "high-risk" technology, not yet mature enough to put in the field. So some artillery experts are promoting an alternative to LP, called a "Unicharge." Unicharge is a new family of NATO propellants that are more efficient and powerful than the normal bagged charges. The use of Unicharge requires a slight redesign of the howitzer (a 1,400-cubic-inch chamber rather than the 1,100-cubic-inch chamber on the M109A6), but it promises the same kind of range and accuracy as the LP weapons currently under development.

So the question is, which one to develop and field? If money were not a consideration, it is probable that the Paladin would be immediately modified to take the Unicharge 155mm gun, and the LP technology would go back to the laboratories for more work. But money *is* an issue, and always will be, especially in these times of drawdown and cutback in the armed forces. While the debate over the risks and advantages of LP versus solid propellants is beyond the scope of this discussion, it is sure to be a volatile and vigorous battle within the artillery community.

On the MLRS side, things are a bit clearer. The XR-M77 and SADARM warheads are ready to be produced; they can be fielded as soon as Congress supplies the money. There is even work beginning on an extended-range variant of ATACMS, to strike up to 186 miles/300 kilometers away. More is to come.

U.S. Army
Aviation Systems

E arly on the morning of January 16th, 1991, an Iraqi technician was walking from one building to another at an air-defense command, control, communications, and intelligence (C^3I) center in south central Iraq. His name and job are still unknown, but on that night he became, briefly, a TV star. For as he walked to the door of the building ahead of him, four bizarre-looking shapes were hovering close to the ground eight kilometers to the south. Suddenly, tongues of fire erupted from the hovering shapes, arching like fiery arrows into the buildings. One after another, the vans, antennas, generators, and barracks of the command center exploded into flame. Stunned by the blasts, the technician started running, probably trying to reach his duty station. He never made it, for as he reached the door to the building and opened it, a Hellfire missile flew directly through the door ahead of him. As the warhead detonated, destroying that side of the building, the blast consumed the technician, making him one of the first Iraqi casualties of Saddam Hussein's insane dream to rule Kuwait. Video cameras attached to the thermal gunsights of the helicopters captured the event in ghostly, blurred images the Pentagon released for public distribution.

The instrument of the technician's death was neither an Air Force F-117A stealth fighter nor a Navy BGM-109 Tomahawk cruise missile, but a flight of U.S. Army AH-64A Apache helicopters. These Army Apaches were landing the first blows of Desert Storm. Even before the high-tech F-117s and Tomahawks hit their targets around Iraq, Task Force Normandy was striking at a pair of Iraqi air defense centers (known as Objectives Nebraska and Oklahoma) along the southern border. The task force was composed of two flights of Apaches (four in each flight; they were code-named Team Red and Team White), along with a pair of Air Force MH-53J Pave Low special-operations helicopters (for communications, navigational, and rescue support if required). The task force was the first element of the Coalition forces to enter enemy territory, and it blasted open two holes in the Iraqi air defense barrier. Destroying these targets was vital, since strike aircraft scheduled to hit SCUD launchers in western Iraq and other targets had to fly through zones controlled by the two air-defense centers. Unless they were destroyed in the opening seconds of the air campaign, there was a real chance that Iraqi air defenses might have inflicted heavy losses on the Coalition aircraft, and warned other targets deeper in Iraq to get ready. The destruction of these targets was so critical that General Norman Schwarzkopf, the CENTCOM commander, demanded a 100% guarantee of success.

The only officer to step up to the challenge of this guarantee was Lieutenant Colonel Richard Cody, an Army Aviation officer of the 101st Air Assault Division. Cody knew that the two air-defense centers had to be destroyed with "to-the-second" timing, and observed to be truly "dead" at the end of the attack. Even the best strike aircraft with the finest targeting systems could not do this; something else was needed. That something was the AH-64A Apache attack helicopter. Lieutenant Colonel Cody felt that the combination of the Apache's firepower, superior thermal-imaging sight, and ability to loiter and observe the results of its attacks made it the only aircraft in CENTCOM capable of the job.

The results of Task Force Normandy's raid that early January morning showed that Cody's confidence in the Apache and the men of the 101st was not just foolhardy bravado. As he and the other fifteen Army soldiers of the task force were firing their missiles, rockets, and cannon shells towards the Iraqis, they were making a statement to the world that Army Aviation had truly come of age. No longer the bastard child of an ugly divorce between the post-World War II Army and its upstart fledgling, the U.S. Air Force, it was ready to take its place as a combat arm for the battlefield commanders of the 1990s and beyond.

The hostilities that led to the divorce between the Army and its Air Corps started after World War I, when flying had a special attraction for visionaries and technological extremists. These men dreamed of mighty bomber fleets that could win a war on the first day with a decisive strike on the enemy's economic and political centers. Army generals on the other hand, by nature a conservative bunch, just wanted a few airplanes with obedient fliers for ground support, which meant strafing enemy trenches, dropping some bombs on command posts and supply dumps, spotting for the artillery, and snapping a few reconnaissance photos, while preventing enemy planes from doing the same.

Clashes over the future role of air power culminated in the court-martial of General Billy Mitchell in 1925. This and other traumatic struggles gave the officers of the Army Air Corps a collective sense of persecution and martyrdom. Unable to secure recognition as a separate service (something the British Royal Air Force had achieved in 1919 and the German Luftwaffe would attain in 1933), the Army Air Corps spent the whole of World War II working hard to prove its strategic importance, all the while lobbying for its own independent branch. This battle it ultimately won, at the cost of thousands of aircraft and fliers lost in heroic daylight bombing missions of sometimes questionable value. With the coming of the atomic bomb, the Army Air Corps had the only means to deliver it, and the birth of a separate Air Force was inevitable. In 1947, President Harry Truman signed the National Security Act, which created the Department of Defense, and made the Air Force an equal branch beside the Army and Navy. The Air Force packed up its planes and bases and left the Army. All it left behind were a few spotting planes, liaison and VIP aircraft, and some wobbly experimental machines called helicopters (which at the time did not seem to have much of a future). With that meager estate, the surviving members of Army Aviation made their new start.

When Igor Sikorsky started delivering helicopters at the end of World War II, they were fragile and unreliable. That was to change in the Korean War, which gave them their first chance for action—as air ambulances. Thousands of United Nations soldiers owed their lives to the noisy "whirlybirds," and a new mission for Army fliers, that of "medevac"—or "dust off"—was born. Throughout the 1950s, advances in engine technology gradually increased the most vital element of rotocraft performance: load-lifting and carrying capacity. During this time, both the Army and the Marine Corps were conducting experiments to find out just what these flying machines could do for them. Meanwhile, both the Air Force and Navy neglected helicopters, concentrating on nuclear-armed bombers, supersonic fighters, and air-to-air guided missiles. In the late 1950s, a breakthrough in technology made the helicopter into a full partner in the world of airpower: the gas-turbine engine. As we have seen earlier, gas turbines (self-contained jet engines that deliver torque to a shaft instead of thrust through a tailpipe) can produce enormous torque with only a minimum of volume and weight. This makes them the ideal power plant for helicopters, which are always trading weight for payload and performance. The new turbines were such an improvement over earlier reciprocating-engine power plants that several designs were converted to take the new engines with only minimal modifications.

With the dawn of the 1960s came the first helicopters designed from the start to take advantage of turbines. The most notable of these was the Bell Model 204, or UH-1, which became famous as the UH-1 Iroquois, or "Huey." Thousands of these versatile choppers were produced and sent to fight in the Vietnam War. So durable was the UH-1 design that new versions and derivatives were still being produced in 1993. One of those offspring was the AH-1 Cobra Attack helicopter, which first saw combat in 1967 and continues in service with the Army and Marine Corps to this day.

These new helicopters allowed the Army and Marines to develop new tactics: For the Marines, who had lost so many thousands of men storming the beaches of Tarawa, Iwo Jima, and other Pacific islands, it was the concept of vertical envelopment—enabling a Marine landing force to come ashore *behind* the enemy without warning. For the Army, it meant the resurrection of the 1st Cavalry Division (which had been disbanded after the Korean War), and making the entire unit moveable by air—or "air mobile"—on a mix of different helicopter types. The new troops—called "air cav"—were the most effective force that fought in Vietnam. Able to swoop down on an enemy without warning, their mobility unimpaired by swamp or jungle, the air cav became a nasty surprise to the Communist forces. This surprise did not come without a price to the Americans, though. The Hueys and other helicopters of their generation had virtually no protection against small-arms fire or the man-portable SAMs that appeared in the late 1960s. Thousands of helicopters were shot down in Vietnam, so many, in fact, that the Army cannot give an exact accounting to this day. And then there was the human cost. The early turbine choppers lacked ballistic or crash protection for the crews or fuel tanks, which resulted in high casualties and terrible burns and injuries to crash survivors. Though fixes and new designs were on the way, they never reached units in Vietnam before the end of the conflict in the early 1970s.

The end of the war in Vietnam and the re-emergence of the Soviet Union and the Cold War as the Army's key focus meant that Army Aviation had to adapt to the new roles and missions of the late 1970s and 1980s. There had been plans to develop a dedicated second-generation attack helicopter to replace the Cobra, but the cancellation of the Lockheed AH-56 Cheyenne program in the early 1970s put an end to these plans. The Cheyenne program suffered from a number of problems, as well as from Air Force complaints that it violated existing agreements on what missions the Army was allowed to fly. The final nail in the coffin of the Cheyenne may have been its very performance. Because of its high speed, and the fact that it derived lift from stub wings, the AH-56 was seen by the Air Force as a technical violation of the Key West agreement. (This agreement, a virtual "treaty" between the Army and Air Force, stated that only the Air Force can own armed fixed-wing aircraft.) But above all, Cheyenne was killed by the changing nature of the threat. Designed to attack from medium altitudes in a steep dive, and protected only against heavy machine-gun fire, it would have fared poorly against the radar-controlled automatic cannon and infrared-homing missiles that the Soviets and their clients were fielding in growing numbers in the late '70s and early '80s.

Nevertheless, like the Armor branch, Army Aviation recovered from the stillborn programs of the 1960s and moved on with improved ideas for the future. Around 1974, they initiated a pair of new programs, the Advanced Attack Helicopter (AAH, which became the AH-64) and the Utility Tactical Transport Aircraft System (UTTAS, which became the UH-60) to replace the venerable but aging AH-1 and UH-1 choppers that made up the bulk of the Army's helicopter fleet at that time. Systems that had been planned for inclusion in the AH-56 attack helicopter, notably the TOW missile system, were grafted onto modified versions of the AH-1, to provide an interim anti-armor capability while the Army waited for new chopper designs to reach service. Finally, the Army began to try out new tactical formations and ideas to make use of its growing aviation assets. These included parceling out aviation brigades to all of the Army's divisions, and not just to dedicated air cavalry formations like the 1st Cav. Also, the aviation community began to integrate their tactics and operational plans with those of the ground forces, so that the overall war-fighting objectives of the Army could more effectively be carried out.

By the early 1980s, these initiatives were beginning to pay off. Army Aviation became a separate branch within the Army in 1983. (Previously, attack helicopter crews belonged to Armor, heavy transport helicopter crews belonged to the Transportation Corps, and scout helicopter crews belonged to the Artillery!) In this way, the Army was saying to its aviators that they were the equals of their Armor, Infantry, and Artillery colleagues. While this may not seem like a big deal, to the aviators it was a special recognition of their combat role. It has also eliminated the kind of infighting that led the Army Air Corps to divorce itself from the Army in 1947. On a more practical level, a whole new generation of helicopter types arrived to enhance the Army Aviation force. Some, like the AH-64A Apache and the OH-58D Kiowa Warrior, opened up whole new capabilities for the Army, like deep-strike and night operations. Others, like the UH-60A Blackhawk and the CH-47D, expanded existing capabilities with improved range and load-carrying capacity. The new helicopters were capable of dishing out awesome firepower and taking punishment that would have destroyed earlier choppers.

Proof of this capability came in Operation Just Cause, the invasion of Panama. And in Operation Desert Storm, Task Force Normandy opened the war, and the mass movement of the 101st Air Assault Division (the only division-sized helicopter unit) to the Euphrates River was one of the war's final acts. So powerful had the Army Aviation units become that General Franks, the commander of the U.S. VII Corps during Desert Storm, had over 800 helicopters under his command.

Let's take a look at some of the tools that have allowed them to compile such an impressive record in the past few years.

The AH-64A Apache Attack Helicopter

It's a nasty-looking beast. I mean, just looking at it, you know you will never see a civilian version as a Traffic-Copter over the city during rush hour. It hardly looks like an aircraft at all. As you walk around it, this monster looks different from every angle. A flying machine, even a helicopter, is supposed to have smooth lines to assist the airflow over its airframe. Not the AH-64 Apache. It has all the direct, in-your-face brutality of some predatory insect spawned on a faraway planet. Except that this one eats tanks, not aphids.

The AH-64 is currently the ultimate expression of the attack helicopter. Its firepower and armor make it the equivalent of a heavy tank flying about the

An AH-64A Apache in your face. Note the windshield wiper on the gunner's front window and the deployed position of the nose-mounted sensors and optics: TADS/PNVS above, laser range finder/designator and direct-view optics below.

McDonnell Douglas Helicopter Company

battlefield, day or night, in adverse weather, finding and killing targets at will, almost immune to enemy weapons.

Like the M1 Abrams, the Apache has its roots in a canceled program. In the case of the AH-64, this was the Lockheed AH-56 Cheyenne. The Cheyenne based its performance more on raw straight-line speed than agility and stealth. Looking for all the world like an angry hummingbird, the AH-56 was designed on the mold of the World War II Russian Il-2 Shturmovik ("storm fighter"), an armored dive-bomber built around tank-killing cannons. Shturmovik's twin 23mm guns could rip through the roof armor of most Nazi Panzers. Some Shturmovik aces racked up hundreds of tank kills.

In addition to the main rotor, the Cheyenne was equipped with a tail-mounted pusher propeller and stubby wings to achieve high speeds (for a rotocraft—over 300 mph/480 kph). The AH-56 was designed to make high-speed dives on its targets, using a combination of TOW missiles (it was the first helicopter designed to fire the TOW system), 2.75"/70mm rockets, and 20mm cannon fire. The performance of the Cheyenne was impressive, but behind the performance lay some fatal problems. For one thing, the cost of the AH-56 escalated during the years of double-digit inflation in the early 1970s. For another, a structural flaw caused the loss of one of the prototypes during testing. Worst of all, the high-speed diving attack of the AH-56 put it in the heart of the weapons envelopes of a number of Soviet-designed weapons (such as the SA-7 Grail and SA-8 Gecko SAMS, and ZSU-23-4 Shilka mobile air-defense gun).

Thus, the stage was set for the cancellation of the Cheyenne. When that happened, there were two major results. First, the Air Force initiated the development of a dedicated close-air-support (CAS) aircraft called Attack-Experimental (shortened to AX), which eventually became the A-10A Thunderbolt II. Second, the Army was given permission to start up a replacement program for the AH-56. That program was known as the Advanced Attack Helicopter (AAH). And this became the AH-64 Apache attack helicopter.

The AAH program was designed to provide the Army with a helicopter capable of day and night, adverse-weather operations against enemy armor and other hardened targets. The Army selected a pair of contractors, Bell Helicopter-Textron of Forth Worth, Texas (the maker of the classic AH-1 Cobra), with their YAH-63 design, and Hughes Helicopter of Culver City, California, and Mesa, Arizona, with their YAH-64, to build prototypes for a competitive "fly-off." Both were excellent designs. The evaluation was long and arduous, and both machines were tested to the limits of their capabilities. In the end, the Army judged the Hughes Helicopters (as of mid-1993 McDonnell Douglas Helicopter Company) entry to be superior in flight performance, cockpit layout, and systems integration. The Army then moved ahead to full-scale development of the Hughes design, now designated the AH-64A Apache. In 1982, the Apache was deemed ready for production, and the first unit was fielded in 1986. The Army has ordered 811 Apaches from McDonnell Douglas Helicopter, with additional units sold to Israel, Egypt, Saudi Arabia, the UAE, and Greece.

The AAH specification made no compromises in the areas of sensors, weapons, agility, and survivability. Unlike the AH-56, where raw speed was the goal, the AAH design emphasized the ability to sneak along at low level, survey the battlefield, sort out the targets, and launch weapons from long range, outside the enemy's anti-aircraft range.

Like the Tank Automotive Command (TACOM) with its common mobility specification for all new vehicle designs, the Army Aviation Center at St. Louis, Missouri, has mandated that all new helicopter designs meet certain standards of maneuverability, ballistic tolerance against enemy gunfire, and load-carrying. For example, the AH-64 is invulnerable to 7.62mm projectiles, tolerant against 12.7mm/.50-caliber projectiles, and survivable (able to get home if hit in the power plant/drivetrain/flight-control systems) against 23mm high-explosive projectiles.

The airframe structure is designed to take a 20-G (twenty times the force of gravity) crash without killing the crew, and the fuel tanks are crash-resistant and self-sealing.

New U.S. helicopters have infrared (IR) signature suppression designed in from the start. Infrared-homing missiles are a major threat to low-flying aircraft. The seeker head on an enemy IR-homing missile is looking for the hot exhaust pipe of a gas-turbine engine. One way to reduce the missile's effectiveness is to mix the hot exhaust gases with a large volume of cooler air, deflect them away from the aircraft, and insulate the exhaust pipes, so that the missile does not "see" hot metal. The AH-64A Apache's "Black Hole" IR suppressors accomplish this very effectively.

Helicopters also need electronic countermeasures (ECM) to survive on the modern battlefield. ECM is a secretive, ever-changing field; and the technical perfomance specifications of particular systems are usually classified, but a typical "suite" of these black boxes includes:

- A radar-warning receiver, to alert the crew when they are being tracked by an enemy radar, so they can take evasive action.
- A radar jammer that transmits signals that drown out or confuse hostile radars.
- Chaff dispensers that release a cloud of metal-coated strips that strongly reflect particular radar frequencies, to clutter up the enemy's radar screen, concealing real targets.
- Flare dispensers that can "decoy" infrared-homing missiles.
- An infrared jammer—typically an electrically heated "brick" on the tail boom of the helicopter that radiates so strongly in a particular IR wavelength that the sensitive seeker head of an incoming missile is saturated and confused. The current model is the ALQ-144, nicknamed "disco ball" because of its distinctive shape.

All this makes the current generation of U.S. helicopters the most survivable in the world. Not invulnerable, but very tough indeed as compared to their

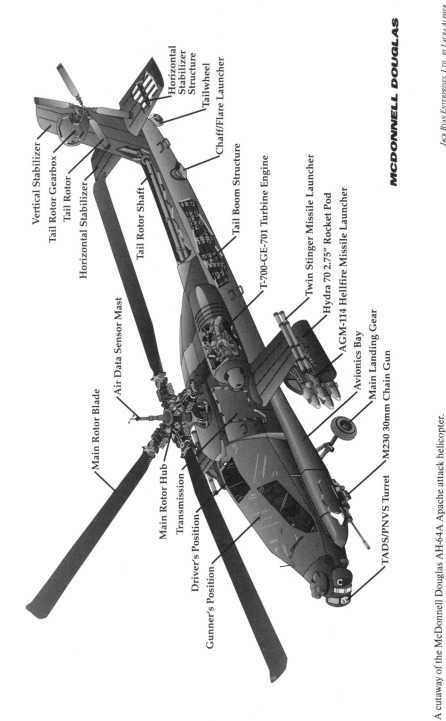

Vertical Stabilizer

Tail Rotor Gearbox

Tail Rotor

Horizontal Stabilizer

Horizontal Stabilizer Structure

Tailwheel

Chaff/Flare Launcher

Tail Rotor Shaft

Tail Boom Structure

T-700-GE-701 Turbine Engine

Twin Stinger Missile Launcher

Hydra 70 2.75" Rocket Pod

AGM-114 Hellfire Missile Launcher

Avionics Bay

Main Landing Gear

M230 30mm Chain Gun

TADS/PNVS Turret

Air Data Sensor Mast

Main Rotor Blade

Main Rotor Hub

Transmission

Driver's Position

Gunner's Position

MCDONNELL DOUGLAS

A cutaway of the McDonnell Douglas AH-64A Apache attack helicopter.

An Army CWO shows series illustrator Laura Alpher one of the avionics bays on an AH-64A Apache attack helicopter. *JOHN D. GRESHAM*

Vietnam-era predecessors. As for load-carrying, the experience of the Southeast Asian jungles has made hot-climate operations a requirement for all new helicopter designs. The magic number "4,000/95" is used as the measure of helicopter performance. This number represents the vertical flight performance that a particular helicopter can achieve while carrying its standard payload in an ambient temperature of 95° Fahrenheit/35°Celsius at a 95% throttle setting. This approximates the worst possible engine conditions (gas turbines generate their greatest horsepower in cold temperatures; their least in hot, humid climates) that might be encountered in places like the Persian Gulf and Panama. Considering the geography of the world's trouble spots, this specification makes good sense.

As you walk around the Apache, you get the feeling that nobody actually designed it, but that a group of guys with blindfolds stuck together a bunch of components with glue and sticky tape. The rotor blades droop, the fuselage sits at an almost absurd upward angle, and stuff juts out of the thing this way and that. But all that is misleading; the Apache is one of the most impressively integrated weapons systems in the world.

The outer skin is mostly aluminum semi-monocoque (i.e., the skin and its underlying ribs are formed into a single load-bearing structure); and much of this consists of access panels. The engine covers are designed to support the weight of service personnel and be used as work platforms. In addition, the entire aircraft is designed to be folded, packed up, and carried in a variety of Air Force transport aircraft.

The two engines are General Electric T-700-GE-701Cs with a rated output of 1,800 shaft horsepower (shp) each. They are coupled into a common main transmission, with the tail rotor being driven by a long shaft running the length of the tail boom. This tail rotor, like those on all conventional single-main-rotor helicopters, is used to counteract the rotational torque of the main rotor to maintain the proper flight attitude. The main rotor head, which is above the transmission, carries a four-bladed main rotor, which is designed to be more efficient than the two-bladed UH-1 and AH-1 designs of the 1960s. More blades give you more lift and a smoother, quieter ride—provided that you have the engine power to drive them at sufficient speed, and the engineering skill to design a rotor head that keeps them balanced, controlled, and firmly attached to the aircraft. Some Russian designs have as many as five or six blades.

In fact, the first real sign that most folks have that a helicopter has four blades instead of two is when the familiar "whomp-whomp" sound of the twin rotor is replaced with an aggressive "growl."

Most of the avionics and other "black boxes" are housed in a pair of fairings along the sides of the forward fuselage, and these provide steps to climb into the cockpit of the Apache. The cockpit itself is separated into forward (for the copilot/gunner) and rear (for the pilot) positions by a thick, bulletproof transparent plate. All of the canopy windows are "flat" panels, which are specially shaped to minimize sun glint (reflections), which can show an enemy the position of the helicopter. The cockpit structure is armored to withstand a direct hit by high-explosive 23mm cannon shells. Both positions in the AH-64 have the standard flight controls (cyclic to control forward/aft pitch, and collective to control power to the main rotor) and displays to pilot the aircraft, though each has instruments for the specific task of each operator.

The most important of these are the readouts for the Martin Marietta Target-Acquisition Designation Sight and Pilot Night-Vision Sensor (TADS/PNVS) system, which is mounted in the nose of the Apache. The PNVS portion of the system is located in the top turret of the system, and is composed of a thermal-imaging sight (similar to the technology on the Abrams and Bradley) whose movement is tied to the movement of the pilot's helmet. The helmet is a remarkable item. Individually fitted and adjusted to each crew member, it allows him or her to aim the aircraft's weapons and sensors with a simple turn of the head.

This system is used whenever the Apache is operated under adverse weather, in heavy fog or dust, or at night. The pilot's view is displayed on a small round screen attached to the helmet that snaps down directly in front of the pilot's right eye, above the cheek. This eyepiece also displays other navigational and fire-control data so that the pilot always has the information needed to get around the battlefield.

The rest of the instruments on the control panel are designed so that under blackout conditions they will not impair the pilot's night vision. Most of them are what are called "strip" indicators, meaning that they show their data as a vertical line, but there are also some circular dials like those you might see on an automobile dashboard. When the Apache was designed, the cockpit layout was considered quite advanced. Next-generation systems (such as the AH-64D Longbow Apache models) will replace most individual instruments with a pair of large multifunction computer-controlled video displays.

The primary navigational system of the Apache is the Litton Attitude-Heading Reference System (AHRS), now standard on most Army helicopters. This inertial reference system works with an ASN-137 Doppler-Velocity Measuring System (a small downward-looking radar which senses the movement of the helicopter over the ground). Over a period of several hours, the AHRS tends to "drift" from an exact positional fix, so most Apaches have a NAVSTAR GPS receiver in the front cockpit that lets the gunner manually punch in corrected data. A modification enabling the AHRS to automatically accept GPS updates will be installed soon.

In the front cockpit are the primary controls for the Apache's weapons systems. While the weapons can be fired from either cockpit, it is the gunner in front who has the job of putting the AH-64's ordnance on target. The weapons are aimed by the TADS/PNVS system, which is housed in the lower part of the nose-sensor turret. It is composed of another FLIR sensor, a daylight television camera, a set of direct-view magnification optics, a laser rangefinder, and a laser designator for targeting of laser-guided ordnance. Much like the pilot in the rear cockpit, the gunner has a helmet-mounted sight with a display eyepiece, which provides a targeting view as well as relevant targeting data. Like the pilot's PNVS, the TADS is slaved to the movement of the gunner's helmet, and is boresighted to what the gunner sees. To engage a target, all the gunner needs to do is select an appropriate weapon, place the "death dot" of the helmet-mounted-sight-targeting reticule on the target, and pull the trigger. The fill-control system does most of the work after that.

The purpose of any weapon is to kill a target, and the AH-64 can destroy almost any class of target that it can detect. In a turret under the chin of the Apache is an M230 30mm chain gun (built by McDonnell Douglas Helicopter). This gun fires a lightweight 30mm round (rather than the heavy 30mm round used in the A-10's GAU-8 cannon), based on the Aden/DEFA ammunition family in use since the early 1950s. It fires a shell called an M789, with a tiny shaped charge warhead which can punch through several inches/centimeters of armor. This means that it can disable a tank from the rear or top, or kill virtually any APC or fighting vehicle in use today (except perhaps the Bradley or the British Warrior). The M789 round also has a fragmenting case, which makes it extremely effective against exposed ground troops. The linkless feed system has a 1,200-round capacity.

The rest of the Apache's weapons are stowed on the two stub wings along the sides of the fuselage. Each wing has two underwing hardpoints for missile and rocket launchers. There are also plans to add a hardpoint on top of the wing for a pair of small air-to-air missiles. On some U.S. Army Apaches, the Stinger missile is used for air-to-air combat. Though nobody had the opportunity to use them during Desert Storm, test dogfights using Stinger and the M230 chain gun indicate that the Apache is equipped to deal with any aircraft that comes within range. This is not to say that AH-64 crews are expected to shoot down high-perfomance jets. But they can kill other helicopters or ground support aircraft, like the Russian SU-25 Frogfoot.

Since the first helicopters were armed, small unguided rockets have been part of their armament. The AH-64 is no exception; it can carry an impressive load of 2.75"/70mm rockets (produced by BEI Defense Systems Company). Known today by their nickname of Hydra-70, these carry everything from a

Control Section

Propulsion Section

Pitch Gyro

Guidance Section

Pneumatic Accumulator

Autopilot Electronics

Yaw/Roll Gyro

Fuze

Battery

Forward Warhead

Spool and Detonation Cord

Main Warhead

Laser Seeker

A cutaway view of the AGM-114F
Hellfire anti-armor missile with its
dual-charge warhead.

Jack Ryan Enterprises, Ltd., by Laura Alpher

10-lb/4.5-kg HE (M151) warhead, to smoke (M264) and illumination (M257) warheads, a submunition warhead (M261), and even a flechette warhead (the M255, packed with small projectiles shaped like carpet nails). Each rocket consists of an MK66 rocket motor, a warhead, and an appropriate fuse (point-detonating, delay, or airburst). Hydra-70s are normally carried in nineteen-round launcher pods. The AH-64 can carry up to four such launchers, though during Desert Storm two were usually carried. During deep-strike operations, an extended-range fuel tank can replace one or more of the rocket pods.

From the very beginning of the Apache's design, the need for a really long-range anti-armor missile was clearly understood. The TOW missile is effective, but the need to trail a guidance wire limits the range to about 3.7 km/2.3 miles, and the firing helicopter must remain stationary while the subsonic TOW flies out to its target. Thus, as part of the AAH program specification, a new anti-armor missile was defined as part of the system package. Rockwell International and Martin Marietta developed and produced the missile, which is type-classified as the AGM-114 Hellfire.

Hellfire is a larger missile than TOW, weighing in at something like 99.6 lb/45.3 kg. Unlike TOW, it is guided by the laser designator of the TADS/PNVS system in the nose of the Apache, which allows it to have a much longer range (in excess of 5 miles/8 kilometers) and much higher speed—supersonic. It also has a considerably larger warhead than TOW-2, with more than 20 lb/9.1 kg of high explosive in the tandem warhead (two shaped charges, one behind the other) of the AGM-114F version. If you are wondering just how much damage such a warhead can do, consider that the earlier AGM-114Cs, with a single-charge warhead, not only penetrated the armor of Iraqi T-72s, but blew them completely apart at the welds!

A Hellfire finds its target because an optical seeker in the missile's nose is programmed to look for the spot of laser light "painted" on the target by the TADS/PVNS of the Apache, the Mast-Mounted Sight of the OH-58D Kiowa Warrior, the GLDS system on a FIST-V, or another laser designator. Even the

LANTIRN laser-targeting pod on the belly of an Air Force F-15E Strike Eagle can be used to designate targets for the Hellfire.

Unlike a lone fighter-bomber dropping a single laser-guided bomb (LGB) on a single target, many Apaches might be firing many Hellfires at different targets on the same battlefield at the same time. Each Hellfire needs to "know" which spot of laser light to attack. So Hellfire (as well as all other current laser-guided weapons) was designed to home in only on a particular laser spot that is pulsing with a particular digital code set by the firing aircraft. This not only solves the problem of keeping multiple missiles on course to the different targets, but it also enables one helicopter or ground observer to target Hellfires being fired from several helicopters.

This allows the firing helicopter to be behind a hill protected from enemy fire (the Hellfire's autopilot can be programmed to arch over intervening terrain), while another with a laser designator "paints" the target from a completely different direction. The OH-58D, with its Mast-Mounted Sight (MMS), can poke the MMS head above a tree or ridgeline and guide Hellfires to their targets without exposing any other portion of the helicopter. Another interesting capability of Hellfire is that an Apache can ripple-fire a salvo of missiles with a short interval (say five seconds) between missile launches. If the Apache gunner has a platoon of three or four tanks sitting next to each other, he can direct the laser onto the first until the first missile hits, and then rapidly shift to the next tank, and the next, until he runs out of missiles or targets. Hellfire can even be used as an air-to-air missile, should the Apache be out of Stingers. If an aircraft like a helicopter gets hit by a Hellfire, it is one dead bird!

During its development, Hellfire has stayed on schedule, and pretty much on budget, with few technical glitches; and only a few modifications have been needed. A dual-mode warhead (to overcome the effects of reactive armor) and a new digital autopilot (to allow the gunner to select a high parabolic or low, terrain-hugging path to the target) were added to the basic -A model to create the AGM-114F variant currently deployed on both the AH-64A Apache and the OH-58D Kiowa Warrior. There are also plans for a millimeter-wave guided version, called Longbow Hellfire, for use later in this decade.

What is it like to actually use this technology? I recently had the opportunity to fly in the front seat of an AH-64A Apache at Fort Hood, Texas, as a guest of the III Corps. My instructor pilot was a chief warrant officer fourth class (CW-4) named Sandy, a lean six-footer who spoke with the vaguely Southwestern drawl that a lot of aviators adopt. From my point of view, the most important part of our preflight conversation was the question "How many hours of stick-time do you have?"

"Oh, about five thousand total," Sandy replied, and then continued. "Twenty-five hund'rd in Snakes [AH-1s], and twenty-five hundred more in the 'Pache."

I knew I could relax. Not always a happy airline flyer, I do like helicopter flying, especially with an experienced CW-4 doing the driving, and Sandy would prove to be as smooth on the stick as the Texan Van Cliburn is on a Steinway. Before my hop that evening, I had the opportunity to get some time in the Apache flight simulator; and I also spent about thirty minutes getting fitted for my flight suit and the helmet-mounted eyepiece sight. Fitting the helmet and sight is not so much difficult as tedious.

As the Texas sun set in the west, Sandy and I headed out to our Apache; and I climbed into the gunner's position and buckled up. The harness is a five-point restraint similar to those used by race car drivers. All that is needed to get locked in is to push each of the buckles into the central harness lock and tighten up the belts. The straps are released by just twisting the buckle knob, and the belts come free. The seat is extremely comfortable. You feel like you're sitting inside a large, transparent greenhouse. After I got the helmet on and the sight adjusted, it was just a matter of waiting for Sandy to finish the preflight routines and get the engines running.

While he did all this, Sandy was kind enough to keep me fully posted on the intercom. The start-up sequence proceeded smoothly, and I was able to monitor it on the instruments in the front cockpit. The only unpleasant moment came when we were backing up from our parking position on the ramp to the taxiway. During that time, exhaust from the turbines was sucked into the air-conditioning system and fed through the vents into the cockpit. Sandy had warned me about this. It only *smells* like a cockpit fire, he told me, but it's quite harmless. He was right. It smelled like the diesel exhaust you get when you stand behind a transit bus. As a matter of fact, the air-conditioning system is not really there for the comfort of the crew, it's for the onboard electronics and instruments. Nevertheless, it's worth the occasional odors, especially when you're wearing a thick Nomex® (fire-resistant) flight suit.

Taking off in a helicopter is almost exactly the same as the feeling you get the moment a mountain cable car lifts out of its barn—an odd vertical lurch followed by a lean forward. In spite of this, you feel pretty safe in the Apache. For my part, my first impression was that I was sitting in an armored bathtub. There is ample protection around you (it's designed to stop impacts of projectiles up to 23mm cannon high-explosive rounds, remember). What with that, and the five-point restraint system, you develop a feeling of security that the simple lap belt on an MD-80 just doesn't provide. (A note to the airlines: Your passengers will stay calmer if they hear what the pilot up front is saying.) Once we had taxied into position and obtained clearance from the tower, we lifted off into the dusky sky, accompanied by two or three other Apaches and a Sikorsky UH-60L Blackhawk carrying several members of my research team. We cruised through the Texas night out to the demonstration area at a smooth 145 knots/265 kph, with the other Apaches and the Blackhawk in trail. The normal ride on a single-rotor helicopter is something like sitting on a chandelier during an earthquake. The aircraft always seems to be vibrating around and from a single point over your head. But because the Apache main rotor has four blades, the ride goes much easier. Once we got to the demonstration area, the first item on the agenda was a demo of what the -64 can do acrobatically. I have to admit that my enthusiasm for this wasn't exactly unqualified, but with my growing confidence in Sandy, I sucked it in and decided to see what my tolerance limits were.

First we went into a hover at about 1,000 feet/305 meters above the countryside. Strangely—I have no idea why—heights always seem less impressive in a helicopter than they do atop a tall building like the Empire State. Sandy then came on the intercom and told me to stand by while he put his bird through its paces. Rapidly, we transitioned to a series of sharp banks, dives, and climbs. Suffice it to say that the AH-64 has the agility to enrage an enemy gunner on the ground. The physical sensation was about like riding the Space Mountain roller coaster at Disney World in Florida. In Sandy's skilled hands, the aircraft hovered, turned, soared up, dove down, accelerated forward and back, and most remarkably, did over fifty knots *sideways.* I have to say that Sandy did all this in such an entertaining way that I was too busy being impressed to turn pale. That agility, of course, is supposed to make life harder on ground gunners and SAM operators and safer for the Apache air crews.

The AH-64 requires a delicate touch, not unlike that of a fighter aircraft. But like a fighter, its sensitive flight controls reward that touch. The precision with which Sandy executed the maneuvers was a message in and of itself. He always told me beforehand what would happen, partially to warn me and partially, I think, so that I could appreciate the skill involved. Having flown the simulator just that afternoon, I did. The aircraft itself radiates a feeling of power and solidity, and an experienced pilot like Sandy only makes it better. Very quickly I was enthralled.

With the preliminaries completed, Sandy headed the Apache off into the Texas night. Because the gunner sits in the front seat of the -64, he surely has the best seat in the house. If you can imagine the view in a particularly well-designed car, then double that sensation and imagine yourself up in the air as well. It's how an eagle sees the world. If you have an aversion to heights, don't worry about it. That doesn't seem to matter if you're sitting down and strapped in. Suddenly you're at home up there. In front of you are some of the most interesting tools you've ever seen. And they're not all that hard to use.

Soon you're no longer an eagle, but an owl. For Apache is most of all a night hunter—except for you as an Apache gunner, it's not night at all. Outside it was clouding up with an impending thunderstorm; things were starting to look like the inside of a cow. To the naked eye there was no moon or stars, but with the thermal-imaging sight every detail on the ground was clear in the green-and-white display. You have a choice of your field of view. And when you spot something interesting, you can zoom in on it by just clicking a button on the TADS controls. At that point your fire-control systems can (at another flick of the controls) lock on the target, and track the target automatically if it moves. This greatly lightens the workload, and allows the gunner to observe and select other targets if desired. The whole point of the Apache, after all, is to make life harder on the enemy by making it easy on the good guys.

As we continued our flight around the Texas countryside, Sandy took the liberty to show me a little Air Combat Maneuvering (ACM) technique on the UH-60L Blackhawk chase helicopter trailing us. As Sandy yanked the big attack chopper around to get onto its tail, I was quickly able to lock up the Blackhawk and track it through the TADS. Sandy told me that a Stinger, a Hellfire, or the M230 chain gun would all have been options at this point had this been a shooting engagement. All I would have done is keep the "death dot" on the other helicopter, and the ordnance of choice would have inevitably impacted it. While there is some additional technique involved when the target is evading, the basics are quite simple to learn and rapidly mastered.

As the storm gathered around us, it was time to head home. During the approach into the field at Fort Hood, the crosswind gusts grew until the trees on the ground were shedding leaves and leaning into the prairie squall that would hit us later that evening. In spite of this, the Apache was steady, and Sandy's control of it authoritative. Landing in the AH-64 is just a simple flare, and then you are on the ground. Before I even knew it, we were rolling back to the parking ramp, where we'd soon talk over the flight with Sandy and the ground crew. All in all, a thoroughly impressive demonstration of the Apache's capabilities. Perhaps most impressive of all, though, was that all the things I did that night, I did at night. Sandy clearly had no qualms about operating at night on the fringes of a thunderstorm with a civilian in the front seat; and that shows a tremendous confidence in the aircraft, as well as in his own experience and skill level. Wherever you are when you read this, Sandy, thanks for the look and the ride!

The Apache is clearly the battle tank of the Army's aerial fleet. Its ability to carry a huge load of different weapons at any time, day or night, and in almost any weather, makes it the weapon of choice for the commander who absolutely, positively has to destroy something hostile. According to Army records, the Apaches deployed during Desert Storm knocked out:

- 837 tanks and tracked vehicles
- 501 wheeled vehicles
- 66 bunkers and radar sites
- 12 helicopters (on the ground)

Two green-screen Multifunction Displays (MFDs) dominate the pilot's station in the AH-64D Longbow Apache. The displays are surrounded by buttons that allow the pilot to choose various options and readout "pages." The small display above the right-hand MFDs is for the navigational systems. Note the small number of analog backup instruments for use in the event of battle damage or power failure.

McDonnell Douglas Helicopter Company

- 10 fighter aircraft (on the ground)
- 120 artillery sites
- 42 SAM and AAA gun sites

In addition to the listed kills, the Apache also assisted in capturing 4,764 Iraqi prisoners.

And this takes us back to where we first started this discussion. For while the Apache is an excellent weapons system today, there is a massive upgrade plan that is being implemented for the future. In late 1996 a new Apache will enter production. Like the M1A2 and Paladin programs, the Army plans to digitize the Apache, with many of the capabilities that advanced ground vehicles have received. Thus, the Army Aviation and Troop Command at St. Louis, Missouri, has given McDonnell Douglas a contract to develop the AH-64D Longbow. Like many of the other new Army systems being fielded as part of General Sullivan's new force, the AH-64D will be re-manufactured from existing AH-64A airframes.

The plan is to strip out all of the existing electronic systems and replace them with new digital systems tied to a 1553 data bus. In addition, all of the cockpit instrumentation will be replaced with a combination of multifunction displays (MFDs) to ease crew workload. This is an important point, because on the AH-64D model, there is going to be a brand-new sensor system called Longbow. Longbow is a mushroom-shaped radar mounted on top of the Apache's rotor mast. The Longbow millimeter-wave radar is designed to see ground and air targets in any weather, day or night. The AH-64D "pops up" from behind masking terrain or trees, and the radar makes just a few sweeps (it can scan either 360° around the Apache or just small pie-shaped sectors). Because the Longbow is designed for stealth, it is hard for enemy sensors to intercept or detect. Once the Longbow has made its sweep, the Apache drops

A prototype AH-64D Longbow banks sharply in flight testing near Mesa, Arizona. Note the mast-mounted radome and the full load of sixteen AGM-114 Hellfire missiles.   *McDonnell Douglas Helicopter Company*

back down under the terrain, and the onboard computer goes to work. Within seconds, the radar's computer processor can detect and classify up to 256 targets by the following five categories:

- tracked ground vehicle
- wheeled ground vehicle
- air-defense vehicle
- airborne fast fixed-wing aircraft
- airborne helicopter

The computer assigns each target a track ID number, and gives it a time/position and a speed and heading fix. Then the data is transmitted to every other airborne AH-64D and other compatible systems tied into the network (like IVIS, MCS, and AFATDS). Thus, the Longbow Apache is functioning as a battle-management platform (like a smaller version of the Air Force J-STARS radar plane or the Navy's Aegis cruisers).

Tied to the development of Longbow has been a new version of Hellfire called Hellfire Longbow. One of the guidance options for Longbow Hellfire is a millimeter-wave seeker that can be programmed to fly to a point over a suspected target, where it switches itself on. The Longbow Hellfire seeker is called a "brilliant" seeker, because it can discriminate between the different types of targets described above. When it sees its assigned target, the missile dives into the target, killing it. While an AH-64A Apache can fire and guide only one Hellfire at a time, the AH-64D can share Longbow radar data and fire up to sixteen Longbow Hellfire in a few seconds. As a bonus, the fire of many Apaches can be coordinated from a single Apache, allowing concentrated massing of fire. A flight of four AH-64Ds, each carrying a full load of sixteen Longbow Hellfires, might destroy up to 128 targets in just a few minutes. Even allowing for misses, it would be like annihilating three or four armored battalions in a few salvoes. Such is the firepower envisioned by General Sullivan: A handful of helicopters can literally demolish a tank brigade with the flick of a switch.

Most models of the Blackhawk helicopter can be fitted with the External Stores Support System (ESSS). The external fuel tanks hold 230 gallons/920 liters each. This version is the MH-60K Special Operations Aircraft, which also carries an air-to-air refueling probe, as well as special night-vision and navigation equipment. *SIKORSKY HELICOPTER-UNITED TECHNOLOGIES*

Currently, the plan is to equip one of every three AH-64Ds with the Longbow radar. Deliveries to the field begin in early 1997, with the first units equipped by late 1997.

The UH-60 Blackhawk Utility Helicopter

One of the hardest jobs in the world is trying to replace a classic. The HMMWV, for instance, had the tough job taking over from the Jeep, which it did—splendidly. And in doing that, it showed how common-sense engineering and mature technologies can make a new classic. This was the job Army Aviation took on when it set out to replace the ubiquitous helicopter of the Vietnam War, the UH-1 Iroquois. The Huey was loved by its crews, and appreciated by all of the military forces that have flown it. Even so, the UH-1 had to be replaced. There was an urgent need for an aircraft with improved ballistic protection, crashworthiness, load-carrying capacity, and survivability.

To obtain this replacement, the Army initiated the Utility Tactical Transport Aircraft System (UTTAS) program in the early 1970s. Three competitors submitted UTTAS program bids (Bell, Boeing-Vertol, and Sikorsky), with Boeing-Vertol and Sikorsky being selected to build prototypes for a competitive "fly off." By 1974, both Boeing-Vertol, with their YUH-61A, and Sikorsky, with the YUH-60A (called the S-70 by Sikorsky), were ready to go head to head for the right to build UTTAS—clearly the biggest post-Vietnam Army Aviation program, both in terms of money and units. By 1976, the competition was completed, and the Sikorsky entry was judged the winner. Christened the UH-60A Blackhawk by the Army, it headed into production in 1979, with Sikorsky and the Army entering into a series of multi-year procurement contracts that still continue. To date, the Army has taken delivery, or has on order, over 1,500 UH-60s and their derivatives. In addition, Sikorsky has delivered hundreds of UH-60 and S-70 derivatives to other branches of the military (the U.S. Navy uses them for antisubmarine warfare and surveillance as the SH-60B/F Seahawk), as well as numerous foreign countries such as Japan, Turkey, and Australia, to name just a few. The UH-60/S-70 airframe has been a huge seller for Sikorsky, and has given birth to a number of different versions, ranging from the basic UH-60L being produced today, to the bizarre-looking MH-60K special-operations variant (one Sikorsky engineer described it as the "Battlestar Galactica") for the 160th Special Operations Aviation Regiment at Fort Campbell, Kentucky.

The basic statistics of the UH-60 family mask the importance of the Blackhawk in and on Army operations. Utilizing a pair of General Electric turboshaft engines, the Blackhawk has a basic weight (dry) of around 10,600 lb/4,818.2 kg, and a maximum gross weight of around 22,000 lb./10,000 kg. The flight crew consists of a pilot, copilot, and crew chief, and there are provisions for carrying eleven fully equipped troops, or fourteen passengers. There are also provisions for carrying a pair of M60 7.62mm machine guns in mounts along the sliding side doors. Top speed is around 160 knots/292.6 kph, with an endurance of roughly 2.3 hours, and a maximum range on internal fuel of 330 miles/603.5 kilometers. Its basic mission as a medium-lift helicopter is to move troops and their equipment into landing zones for air assaults and to haul supplies into places ground transport cannot reach. In addition, the UH-60 has been used extensively as a medical-evacuation helicopter, though a dedicated "dust off" version has yet to be procured by the Army. The UH-60 is designed for rapid deployment by air, with a C-130 able to hold one Blackhawk, and a C-5 able to hold six.

The UH-60 can carry an External Stores Support System (ESSS), which consists of a pair of stub wings (on each side of the fuselage), with plumbing and wiring for extra fuel tanks, cargo pods, "Volcano" anti-armor mine dispensers, or even four-round Hellfire missile launchers. While the Blackhawk cannot fire Hellfires by itself (it lacks a laser target designator), it can *launch* missiles at targets "painted" by any other system equipped with a target designator.

The L-model of the UH-60 is quickly becoming the Army Aviation equivalent of the HMMWV. It can lift the majority of the equipment found in light, airborne, or air assault divisions. Thus, the UH-60L is regarded as a "divisional" lift asset, reducing the burden on the Army's limited supply of CH-47 heavy lift choppers. As the name implies, the UH-60L is a follow-on to the basic UH-60A, with a number of modifications. These include:

- Improved T-701 engines, with 1,940 shp per engine. In addition, the transmission has been up-rated to accommodate the additional horsepower.
- Minor structural improvements have been made to the airframe and external cargo hook to accommodate slung payloads up to 9,000 lb/4,090 kg. This allows the UH-60L to externally carry payloads up to the size of an M1097 "Heavy Hummer" HMMWV.
- The flight-control system has been modified to be more resistant to electromagnetic interference (EMI) that can occur from flying around power lines and other types of heavy-duty electrical equipment.
- The cockpit instrumentation layout has been redesigned to reduce crew workload, particularly during night operations while utilizing night-vision goggles (NVGs). In particular, all of the cockpit lighting has been revised so that it is compatible with NVG usage.

All of this makes the -L model Blackhawk the world's best medium-transport helicopter. Only the Bell Boeing V-22 Osprey being readied for production for the Marines is in the same class, and it is years from entering service.

So what is the UH-60L like in the air? It may be the easiest and most comfortable helicopter to fly in all of the U.S. inventory. As you sit down in the right seat (the pilot sits in the right seat, the opposite of fixed-wing aircraft, though like most two-seater aircraft, the Blackhawk can be flown from either position) and adjust the seat to a comfortable position, the first thing that strikes you is the logic of all the instruments. Most of the important ones are of the "strip" variety, meaning that they are like electronic thermometers, showing their information by the rise and fall of an illuminated light along a scale. In addition, the pilot and copilot each have a block of warning indicators, called enunciators, to show critical information like fire warnings, high temperatures, and landing-gear status. There are the usual aircraft instruments, like an artificial horizon, as well as the readouts for the AHRS, TACAN, and GPS receiver. Much like the AH-64, the UH-60L has a full ECM suite with an RWR, as well as provisions for jammers and decoy launchers. The actual flying controls are conventional (for a helicopter) with the collective (essentially the engine power control) on the left and the cyclic (the pitch and directional controls) on a stick between your legs. There are also foot pedals to help in turning the helicopter during hovers and such, but I found that these are used only occasionally by Blackhawk crews. Once you have strapped in and are ready to get started, the crew chief gets out in front of the Blackhawk (where he's tied in to the helicopter's intercom system by a cable) to watch for any fires or problems during start-up. Engine start is just a matter of pressing a couple of buttons and waiting for the T-701s to warm up. As soon as all the warning enunciator lights are green (red ones are "bad," as you might imagine), then you unlock the parking brake, push the cyclic forward, pull up slightly on the collective to apply power, and taxi forward to your take-off spot. Once you have called the tower for permission to take-off (the Blackhawk is equipped with the new SINCGARS jam/interception-resistant radios), you pull up gently on the collective, pitch the cyclic stick forward, and you're airborne.

Almost immediately, you notice how smooth and silky the Blackhawk flies. Part of this comes from just how well designed and balanced the power plants, transmission, and rotor system are. In fact, I was surprised to find that the Blackhawk does not have active vibration suppression like the OH-58D and some other helicopters. The other part of what makes the UH-60L so smooth is the auto-stabilization system that is tied to the flight-control system. This system smoothes out the control inputs and aircraft responses, and makes the Blackhawk ride like an American luxury car. It is a joy to fly, and the feeling of control, authority, and smoothness immediately translates to confidence in the UH-60's ability to react to any situation. In fact, so easy is the Blackhawk to fly that commercial operators of other types of helicopters find that UH-60/S-70 flight crews frequently have to be retrained to fly what they call "real" helicopters. Hovering is almost simple: All you have to do is pull back the cyclic a bit to generate a small nose-up "flare" (this slows the helicopter down), adjust the collective, and there you are hanging in the middle of open air! It is an amazing feeling.

All of this thrashing about in the air is fun, but low-level flying is what Army Aviation is all about, and the Blackhawk is well qualified for this kind of work. We recently got to join a UH-60L crew of the 4th Air Cavalry Squadron, 3rd Armored Cavalry, at Fort Bliss, Texas, on a hop out to an exercise area. Just to make things interesting, the entire flight was done at night (actually about 3 AM), with the crew utilizing their new AN/PVS-6 NVGs. As we lifted off, the crew was careful to establish their reference points and the line of the horizon. One of the risks in night flying is the sudden onset of vertigo, which can make aircrews trust their eyes and not their instruments. To reduce this risk, the pilot and copilot split up the flying and instrument-monitoring tasks. Periodically they like to switch, so that neither becomes fixated on what he is doing.

As we flew out to the objective, the crew began to fly what is known as a "contour" flight profile. This has the Blackhawk transiting as quickly as possible (doing about 150 knots/275 kph), maintaining a constant altitude above ground—whatever the terrain—of about 50 feet/15.25 meters. Contour flying is an incredible thrill in the UH-60L, and the auto-stabilization system really smooths out the ride. The vertical movements are rapid, though never panicked or sudden. This makes the Blackhawk difficult to track for enemy gunners or SAM operators on the ground. It also tends to discourage any enemy attack helicopter crews or fighter planes from trying to get in a "cheap shot" at the transiting UH-60L. Unless your auto-stabilization system is better than the Blackhawk's, maneuvering into a good close firing position on its tail could cause your flight path to intersect the terrain, terminating your flying career with extreme suddenness.

On the ride out to the landing zone (LZ), the crew demonstrated how they make combat insertions of A-teams (scout patrols) and special-operations personnel. First came several fake insertions: The helicopter touched down and took off without unloading anything. In this way, should someone observe the Blackhawk's maneuvers, they would not be sure of the team's actual LZ. When the time for the insertion finally came, the crew chief told everyone to hold on, and get ready for a rapid stop. The pilot then applied a hard nose-up flare and dropped power, so that the forward motion of the Blackhawk stopped at the same time as the wheels touched the ground. As soon as this happened, the crew chief threw open one of the sliding side doors, the landing team leaped out, and the crew chief rapidly passed out equipment and supplies. Within ten to fifteen seconds, the team was clear, and the crew chief was slamming shut the side door and telling the pilot they were clear to lift off. Up in the cockpit, the pilot applied maximum collective power and pitched the cyclic stick forward to clear the LZ as quickly as possible. The flight crew then made several additional decoy landings before they dropped us off at the exercise area and returned to base.

As the Blackhawk enters its second decade of military service, it has a track record of reliable performance and durability. And Blackhawks continue to be delivered in a variety of different models. In fact, the Sikorsky production line in Stratford, Connecticut, remains the busiest helicopter production line in the world today. The UH-60L will remain the standard medium-utility helicopter for the U.S. Army well into the 21st century.

The OH-58D Kiowa Warrior
Scout Attack Helicopter

Question: What military aircraft has won the largest number of industry and government awards in the past few years for engineering excellence, customer satisfaction, and combat performance in the field? The AH-64 Apache? The F-117A stealth fighter? The OH-58D Kiowa Warrior?

Answer: The OH-58D Kiowa Warrior.

"Huh? The what?" you ask (not surprisingly). Yet even if you've never heard of a Kiowa Warrior, the fact remains that there is no aircraft program anywhere in the U.S. military that has been more successful. What makes this more surprising is that not one new OH-58D airframe has ever been built for the U.S. Army. Every unit delivered for U.S. service has been converted from an existing airframe. If this sounds as intriguing to you as it did to me when I first heard the Kiowa Warrior story, then read on.

Scouting and observation are among the most important of Army Aviation's many missions. In fact, the first military mission for aircraft was airborne spotting by the balloons of Dr. Thadious Lowe during the American Civil War. Later, during World War I, the use of observation/reconnaissance aircraft led to the development of fighter aircraft to shoot them down.

Like so many other weapons we have looked at, the OH-58D story begins in the Vietnam era. During the Vietnam War, the Army acquired a number of small scout helicopters to lead air cavalry assaults and spot targets for attack helicopters. Though the Army initially used the Hughes Helicopter OH-6 (now evolved into the McDonnell Douglas MD-500 series) for this mission, they eventually settled on a version of the popular Bell Helicopter-Textron Model 206 Jet Ranger to provide this vital service. Most people are familiar with the 206 as the helicopter used for traffic and television news reporting. The Army bought a number of these in the late 1960s and early 1970s, installed military radios and avionics (and little else, it turned out), and named them OH-58 Kiowas.

The OH-58 proved adequate for scouting in daylight with the naked eye, but had severe limitations in darkness, fog, or haze. This became a serious problem when the OH-58s had to spot for the new anti-armor versions of the Cobra that were beginning to appear in the 1970s. OH-58 crews were able to sight "something" in the distance, but then they would have to call the Cobra attack helicopters (equipped with long-range stabilized optical systems) they were supposed to scout for so the Cobras could identify the targets for them!

The shortcomings of the OH-58 were well known to the leadership in Army Aviation; but they had to wait until the contracts for the Apache and the Blackhawk had been let before they could afford to shoehorn the scout program into the budget. By the late 1970s, a plan to upgrade the Army's aeroscouts had been formulated under the name of Army Helicopter Improvement Program (AHIP). Under the AHIP plan, the winner of a competition between Bell Helicopter-Textron and Hughes Helicopter (now McDonnell Douglas

Helicopter) would rebuild existing scout helicopter airframes (to keep down costs) with new engines, avionics, and sensors. In 1981, Bell won with their proposal to rebuild the Army's fleet of OH-58 Kiowa airframes into the AHIP helicopters. The key to the new helicopter's capability was the McDonnell Douglas Mast-Mounted Sight (MMS) equipped with a stabilized FLIR, a daylight television camera, and a laser rangefinder and designator for Hellfire and other laser-guided ordnance. By 1985, the first unit of AHIP helicopters was ready to be fielded. And by the time of REFORGER-87, the AHIP Kiowas had made a powerful impression; and commanders from all kinds of units, from field artillery brigades to Apache attack helicopter battalions, were scrambling to get some.

That might have been the extent of the OH-58D story had it not been for those *other* bad boys of the Persian Gulf, the Iranians. Towards the end of their eight-year war, both Iran and Iraq launched a campaign of attacks against the tankers carrying each other's crude oil out of the Persian Gulf. As long as the two belligerents were pumping Exocet guided missiles and rockets into the tankers of their opponents, nobody but Lloyds of London cared. But when the Iranians started attacking tankers servicing the oil trade of other members of the Gulf Cooperation Council (like Kuwait, Saudi Arabia, and the United Arab Emirates, who at the time were funding Iraq's war with Iran), that was another story entirely. And in 1987, the government of Kuwait requested the support of the United States to maintain free passage of maritime trade in the waters of the Persian Gulf.

Shortly afterwards, the U.S. Navy began a long-term operation to escort Kuwaiti tankers (reflagged under American ownership) in and out of the Gulf. Initially, the naval planners thought that Iranian Silkworm missiles and fighter-bombers would be the primary threat to the tankers. This turned out to be a misconception. The misconception was shattered on the very first convoy, when the tanker *Bridgeton* struck a primitive, though highly effective, contact mine laid by the Iranians. Within a matter of days, the Iranians were driving the American forces crazy with a guerrilla-style naval war fought with mines and with Iranian Revolutionary Guards in speedboats (called "boghammers") firing rocket-propelled grenades. The U.S. Navy of the 1980s was designed for open-ocean combat with the Russian Navy, not inshore operations against a low-tech, hit-and-run force that did not play by any set of known rules. The Navy needed help; and (would you believe?) they had to ask for it from the U.S. Army.

After the failed Iranian hostage rescue of 1980, the Army had built up its special-operations capabilities. A special helicopter unit, Task Force (TF) 160 (now known as the 160th Aviation Regiment), was created at Fort Campbell, Kentucky. TF-160 (who call themselves the *Nightstalkers*; their motto is, "We own the night") had a number of modified McDonnell Douglas H-6s (designated AH-6s), equipped with thermal-imaging sights, machine guns, and rockets—just the thing for the Persian Gulf, it was thought. And soon, operating off U.S. Navy frigates and anchored barges, they became known as the "Killer Eggs." The AH-6s turned the tanker war around—their crowning achievement

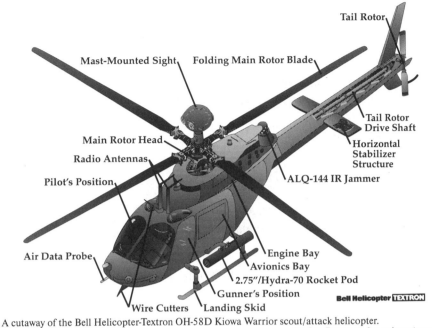

Tail Rotor

Mast-Mounted Sight   Folding Main Rotor Blade

Tail Rotor
Drive Shaft

Main Rotor Head

Horizontal
Stabilizer
Structure

Radio Antennas

ALQ-144 IR Jammer

Pilot's Position

Air Data Probe

Engine Bay
Avionics Bay
2.75"/Hydra-70 Rocket Pod
Gunner's Position
Wire Cutters   Landing Skid

Bell Helicopter TEXTRON

A cutaway of the Bell Helicopter-Textron OH-58D Kiowa Warrior scout/attack helicopter.

JACK RYAN ENTERPRISES, LTD., BY LAURA ALPHER

being the capture (and destruction) of the *Iran Ajar*, a landing craft that was caught laying mines.

Unfortunately, because it had to carry out other urgent commitments, the Army recalled its tiny fleet of AH-6s (and their special-ops crews). At the same time, the Army realized that an aircraft was needed to replace the AH-6s in the Persian Gulf, one that could be operated by regular Army Aviation troops.

In a little while they concluded that an OH-58D, modified to fire air-to-surface weapons, would do very nicely. In September of 1987, under a "black" program (the very existence of the program was secret) code-named PRIME CHANCE, the Joint Chiefs of Staff directed Bell Helicopter-Textron to convert fifteen OH-58Ds to an armed configuration. Completing prototyping, testing, and fabrication, the contractor delivered the first two aircraft to the Army less than one hundred days after go-ahead. Within seven months, fifteen PRIME CHANCE aircraft were delivered to the 1st Battalion of the 18th Aviation Brigade (assigned to XVIII Airborne Corps). The modifications to the basic OH-58D included:

- The installation of weapons pylons capable of taking AGM-114 Hellfire missiles, air-to-air Stingers, 2.75" Hydra-70 rockets, and a .50-caliber machine-gun pod.
- Uprating of the maximum continuous power of the engines and transmission from 455 to 510 shp, as well as use of a different lube oil to handle the high temperatures of the Persian Gulf.

- Installation of a mission equipment package consisting of an ARN-118 TACAN navigational receiver, a video recorder for use with the MMS, and some new avionics (an MIL STD 1553 data bus was already standard on the aircraft).
- An electronic-countermeasure suite consisting of an AN/APR-39/44 RWR, and AN/ALQ-144 IR jammer.
- Ladders for overwater crew rescue.
- Shielding for the expected electromagnetic interference from the radars of the Navy ships that TF-118 would operate from.

Operated under the code name of TF-118, the PRIME CHANCE OH-58Ds rapidly swept the Persian Gulf of the Iranian forces harassing the tanker trade. After only a few engagements, the Iranian boghammers and mining vessels apparently decided not to mix it up anymore with the TF-118/Navy team. Shortly afterwards, the Iranians and Iraqis reverted to firing SCUD missiles on each other's cities—while negotiating an armistice. In 1989, as the last convoy escorted by the U.S. Navy left the Persian Gulf, the last U.S. asset in the Gulf was an airborne PRIME CHANCE OH-58D watching the back door. So effective was the performance of the PRIME CHANCE OH-58Ds, in early 1990 the Secretary of the Army ordered that all 243 of the Army's OH-58Ds be armed like the PRIME CHANCE aircraft, and that another eighty-one more be bought to meet the demand for the little helicopter with the big eye.

The dust from the Kuwait tanker escort operation had hardly settled when Saddam Hussein's forces invaded Kuwait in August of 1990. Immediately, the fifteen PRIME CHANCE aircraft (now assigned to the 4th Squadron of the 17th Cavalry, XVIII Airborne Corps), as well as the rest of the OH-58D population, were sent to the Persian Gulf to serve in Operations Desert Shield and Desert Storm. The PRIME CHANCE birds reverted to their nautical haunts of the tanker war, operating off barges and Oliver Hazard Perry-class (FFG-7) guided-missile frigates. Of particular note was the service of a pair of PRIME CHANCE OH-58s assigned to operate off of the USS *Nicholas* (FFG-47). Starting in late January 1991, the two Kiowas did everything from armed reconnaissance, to destroying a Silkworm missile site, to sinking Iraqi patrol boats. The record of their landlocked AHIP cousins was equally distinguished: scouting and spotting for the Apaches, "painting" laser spots on targets for Copperhead rounds, night scouting along the front, and target-spotting/handoff to fixed-wing aircraft. Through all of their Persian Gulf operations, from 1988 to the end of Desert Storm, not one OH-58D was lost to enemy fire. In recognition of the growing reputation of the armed OH-58D, it was renamed the Kiowa Warrior, to reflect its achievements in a short and adventurous career.

What is the production OH-58D Kiowa Warrior all about? As you walk up to one, your first thought might well be that the sleek lines of the original Model 206 have been ruined by all the antennas, wire cutters, and that awkward ball at the top of the rotor head, the Mast-Mounted Sight (MMS). But of course, all that

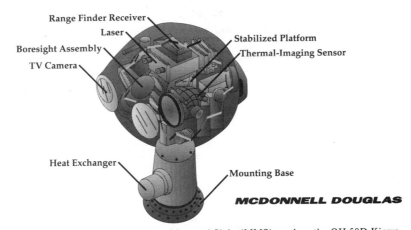

Range Finder Receiver
Laser
Boresight Assembly
TV Camera

Stabilized Platform
Thermal-Imaging Sensor

Heat Exchanger

Mounting Base

MCDONNELL DOUGLAS

A cutaway of the McDonnell Douglas Mast-Mounted Sight (MMS) used on the OH-58D Kiowa Warrior scout/attack helicopter.   *Jack Ryan Enterprises, Ltd., by Laura Alpher*

stuff (and some other stuff) is what gives the Kiowa Warrior its special character.

The MMS itself sits on a special vibration-isolation mount above the four-bladed rotor head. Within the ball-shaped housing are a FLIR thermal-imaging system, a daylight television system, a laser rangefinder, and a laser designator. These systems are tied into the cockpit through a pair of multifunction displays (MFDs). Resembling a miniature 6-inch/15-centimeter green computer screen, the MFD is surrounded by fourteen buttons that bring up various menus and screens (called "pages" in crew jargon) that control and monitor the MMS, navigation, and other flight systems. Depending on the preferences of the users (both flight crew members have one), the MFD can be used as a visual/targeting display for the MMS, to display navigational waypoints, or monitor fuel and weapons status. All of the controls for the weapons and the MMS are located on the controls (cyclic and collective grips), and either crew member can fly the aircraft, with the copilot operating the MMS and the pilot firing the weapons. For example, the controls on the pilot's cyclic grip include such things as weapons select-and-fire buttons, MMS controls, trim controls, and MFD controls.

The cockpit itself shows its light helicopter lineage: It is a tight squeeze for a big passenger. The middle cabin of the OH-58D has been turned into an avionics bay, and the flight crew (pilot in the right seat, copilot on the left, both with controls) are seated in the front cabin. The seats, while comfortable, are cramped if you are over 6 feet/1.8 meters tall. There is a small heads-up display (HUD) for the pilot. To assist in mission planning and navigation, there is a data loader similar to that in the Apache, so that missions can be set up on a personal computer back at base, then transferred to the helicopter on a memory unit resembling a video game cartridge. In addition, a video recorder has been installed to record all of the information from the MMS cameras and data from the display pages.

One shortcoming of the Kiowa Warrior is that there is no way to completely seal the cockpit against chemical attack. And in fact, on hot days, the cockpit doors are often removed. To provide for operations under chemical contamination, the crew wears MOPP-IV chemical suits; and there is an onboard system to feed filtered air through the M-43 aviator's face mask. For night operations, the crew has to use a set of AN/PVS-6 low-light goggles, which are clipped on over their helmets. This provides a limited field of view, somewhat like looking down a cardboard tube. The cockpit instruments and displays have been designed to be usable by the crew when they are in goggles.

The weapons are carried on a pair of tubular metal pylons attached amidships on the fuselage. A .50-caliber machine-gun pod can be mounted on the left pylon only. Each pylon can carry any of the following:

- two AGM-114 Hellfire missiles
- two air-to-air Stinger missiles
- one M260 seven-round 2.75"/Hydra-70 rocket-launcher pod

Any combination of weapons can be carried, though normally the Kiowa Warriors operate in pairs with one aircraft carrying Hellfires (two) and rockets (one pod), and the other loaded with Stingers (two) and a machine-gun pod. Of course, with the MMS, the Kiowa Warrior can designate a target for any kind of laser-guided ordnance. In addition, there is an Airborne Target Handover System (ATHS) which can transmit the coordinates of a ground target by digital data burst, as fixed by the MMS sensors, to any number of aircraft such as Army AH-64s, Air Force A-10 Warthogs, and Marine AV-8B Harriers. There are also direct connections for the TACFIRE artillery-control system (and the AFATDS when it arrives).

All of these systems, as well as the crew's voice communications, are fed through a pair of radios (a VHF AN/ARC-186 and a UHF HaveQuick II), which can be controlled through the MFDs. There even is an option the Army has installed on a few aircraft for a real-time video downlink from the MMS to commanders on the ground. Another option is something called a "night pilotage system," which would involve installation of a small thermal-imaging sight in a turret under the nose. This would function like the PNVS system on the Apache, feeding data to the crew through helmet sights. But for now, budget restrictions will keep these options on hold.

Flying the Kiowa Warrior is different from the other helicopters we have talked about. Where flying the Apache involves the subtle use of power, and the Blackhawk rides like a friendly Cadillac, the Kiowa Warrior is more like a little imported sports car. Strapping in is similar to getting into the Apache or the Blackhawk.

Since OH-58D crews often fly with the doors off, there is a lot of noise; and it is vital to have a helmet and flight suit that fit comfortably. When the preflight checks are completed, the pilot just pulls back on the collective, and you're off. The agility of the Kiowa Warrior is amazing, though it can only do about 120 knots/219.5 kph with the doors on (you lose about 10 knots/18 kph flying with the doors off). But unlike the Apache or the Blackhawk, high dash-speeds are not what the OH-58D is about. Being sneaky is the OH-58D's strong suit. The Kiowa Warrior's small size and agility enable it to slip between tree lines, or down a streambed, to sneak up on an opponent. With only the MMS poking above a line of trees or a ridge, the scout is almost invisible; and the four-bladed rotor reduces the blade noise (outside the cockpit). Gone are the days when a chopper like the UH-1 would announce its presence for miles with a distinctive "whump-whump" from its two-bladed rotor. As the crew surveys the scene with the MMS, they can report back to headquarters by voice radio, directly to other aircraft through the ATHS, or to artillery fire units via TACFIRE.

Normally, the OH-58D will be the eyes for other systems that shoot. But if necessary, it can be a very dangerous shooter itself. Its maneuverability, particularly down low in terrain, means that very few weapons can track or maneuver with it. Setting up a shot with Hellfire, the MMS, and the sensor controls is almost absurdly easy. As an example, my researcher (no pinball wizard) was able to lock on to a window on the top floor of a hotel some 6 miles/10 kilometers away with the MMS; and as the aircraft maneuvered he had no problem keeping the target centered in the sight. Launch of an actual missile would have placed it squarely through that window. And the Kiowa Warrior may be the best existing helicopter for dogfighting other helicopters. While the air-to-air Stinger is a big part of this, maneuverability is the key. Believe it or not, helicopters with four-bladed rigid rotor systems can both roll and loop, and the Kiowa Warrior is quite capable of aerobatics. If anything, the Kiowa Warrior is overpowered and overly sensitive, requiring the calming hand of a thoroughbred jockey to get the most out of it.

Today, as the armed OH-58Ds roll off the conversion/assembly line at Bell Helicopter's plant in Fort Worth, Texas, they represent one of the best bargains in the U.S. arsenal. Some fifteen have even been converted into a "low observable" configuration, and operated by the 1st Squadron of the 17th Cavalry. These special Kiowa Warriors, simply known as "upgrade" birds, have redesigned noses to reduce their radar cross-section, as well as radar-absorbent material (RAM) on the side doors, the MMS, the main rotor head, and tail rotor. There is even a special gold-colored windshield coating that apparently has radar absorption qualities. This means that the "1st of the 17th" has fifteen stealthy scout/attack helicopters, capable of going God-knows-where to do God-knows-what. One can only speculate.

The Army would like Bell to convert a total of some 507 Kiowa Warriors, though only about 366 have been contracted to date—which brings us to a financial dilemma. A newly converted OH-58D coming off the line costs about five million dollars, as opposed to perhaps twelve million dollars for one of the new RAH-66 Comanche scout/attack helicopters soon to come on-line. Which makes Kiowa Warrior look like quite a bargain. The problem with such a comparison is that the OH-58D airframe has limited growth potential, while the RAH-66 design is at the very beginning of its life cycle. What this means is that with the OH-58D you must give careful consideration to adding things to the old

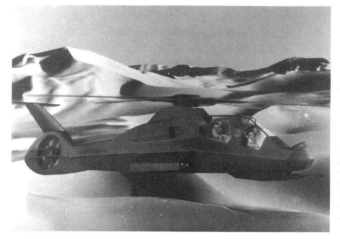

A view of the RAH-66 Comanche. Note the enclosed FANTAIL® tail rotor, and the recessed low-observable engine air intakes. The 20mm gun barrel can pivot down and to the rear for stowage and the missiles are mounted on doors that retract flush with the skin of the aircraft.
BOEING-SIKORSKY

Model 206 airframe, while the future of the new RAH-66 is wide open. In other words, as long as buying the OH-58D does not conflict with the Comanche program budget, the Army will probably support contracting for the 507 aircraft in the current requirement. But whether they do that or not, they have still managed to acquire the best light scout/attack helicopter in the world today.

The RAH-66 Comanche Scout/Attack Helicopter

Comanche is the top-priority program in the Army. Not just Army Aviation, mind you, but the whole U.S. Army. If you have any doubt of this, consider that the gentleman who told me was the Army Chief of Staff himself. I learned this during a briefing in General Sullivan's office last summer, as he outlined his plans for the Army of the 21st century. Now, when an old cavalry officer tells you that a *helicopter* program is the most important thing the Army is buying, it is worth taking a closer look at that program.

Up in Stratford, Connecticut, in Philadelphia, and around the country, the combined efforts of Boeing Helicopter, Sikorsky Aircraft, and a team of other contractors have just begun assembling composite fuselage parts and integrating electronics for the most important advance in helicopter technology since the turboshaft engine, the RAH-66 Comanche scout/attack helicopter. Comanche is designed to replace the Army's aging fleet of AH-1, OH-6, and OH-58 helicopters with a single airframe capable of handling all scout/light-attack missions. And when it gets into the force in 2003, it is going to change the face of combat forever.

Why?

For one thing, the RAH-66 will be as nearly invisible to radar, audio, and infrared detection as modern technology can make it. For another, it is being engineered to be the most capable and survivable sensor system in the Army. But most important, it is designed to get the answer to the ground commander's eternal question, "What's on the other side of that hill?" Recall that General Franks had to commit the whole of VII Corps to combat with the Republican Guards on the basis of very limited information obtained in bad weather by a handful of young cavalry officers. Comanche is designed to cut through that "fog of battle." It will use a combination of advanced sensors and electronics, stealth technology, and high maneuverability to find the enemy while itself remaining unseen.

Even before Desert Storm, the Army knew it needed such a helicopter. The replacement program, known as Light Helicopter-Experimental (LHX), was designed to fill in a single airframe the requirement for both a new scout helicopter and a light-attack helicopter. The Army Aviation Command at St. Louis, Missouri, selected two "superteams" of top contractors to design "paper" aircraft to compete for the final full-scale development contract. One team was led by McDonnell Douglas Helicopter and Bell Helicopter-Textron, while the other had Sikorsky and Boeing Helicopter as its principals. The competition was fierce, since this was clearly going to be the last big military helicopter contract of the 1990s, and orders for existing models were already dropping.

Both designs had two-man crews and employed stealth technology. The McDonnell Douglas/Bell design eliminated the tail rotor by using a ducted fan (called NOTAR, which stands for No TAil Rotor). The Boeing®-Sikorsky team used a ducted tail rotor called FANTAIL®.

In the end, the Boeing-Sikorsky team was selected to build the prototypes and to take the new helicopter, designated RAH-66 Comanche, into production. In 1993, the first structural components for the flying prototype were produced. Current plans put the first flight of the Comanche sometime in 1995, with the first unit to be fielded in 2003. The Army wants to buy some 1,300 RAH-66s to replace over 3,000 AH-1s, O/AH-6s, and OH-58A/C/Ds currently in service.

So just what will this new helicopter have in the way of features? Well, consider the following list:

- Crew—The Comanche will carry a two-man crew in the same kind of tandem arrangement as the AH-64. The cockpits will be fully integrated with an array of programmable MFDs, which can be rapidly reconfigured in flight by the crew. In addition, the crew helmets and sights will be of a new lightweight design that will eliminate the bulky versions found on the Apache and other U.S. helicopters. Also, the whole compartment is ballistically protected and sealed with an environmental filtration/overpressure system, for protection against nuclear, biological, and chemical contamination.
- Structure—The majority of the RAH-66 will be composed of fiber, carbon, and plastic composites. Metal components will be minimized, for both stealth and weight considerations. In addition, the Comanche is being built to the same standards of

ballistic protection/tolerance as the AH-64 Apache and UH-60 Blackhawk.

- Engines—A new company, LHTEC (a joint venture by Garrett and Allison) of St. Louis, Missouri, will produce the uprated T-800 engines for the Comanche. The engine inlets are buried to reduce their radar signature, and the exhausts are cleverly concealed in the tail boom, where the hot gases are mixed with cooler ambient air and vented downward to reduce the Comanche's IR signature. Rated at 1,380 shp each, the engines will probably be the standard power plants for all new U.S. light and medium helicopters well into the 21st century.

- Sensors—Comanche will carry a targeting and piloting system similar to the TADS/PNVS on the AH-64A. What will make the system on the RAH-66 different is that the thermal-imaging system will use "second-generation FLIR" technology. This means that the imaging elements have a much higher resolution and sensitivity than those currently deployed on U.S. weapons systems. The thermal pictures will be good enough for the onboard computer to positively identify a target (airborne or ground) solely from its thermal signature. This means that the system will automatically be able to tell the difference between an American M1A2 and a Soviet-designed T-72. In addition, all Comanches will be wired for installation of a Longbow radar on top of the rotor head. As with the AH-64C/D, the Comanche will have a modem to transmit data to other users on the network. About one RAH-66B (with Longbow) will be fielded for every three RAH-66As (without Longbow). Comanche will also carry a laser target designator and a full electronic-warfare suite, including radar-warning (RWR) and (eventually) jammer equipment.

- Flight Avionics—The flight avionics on Comanche will be the most comprehensive of any helicopter flying, including the Air Force Pave Low special-operations birds. Along with the now-standard GPS receiver, AHRS, SINCGARS radios, and other navigational gear, there will be a moving map display right out of a James Bond movie to help the crew maintain their bearings, manage the battle, and pass along information to the rest of the force. Previously, the only aircraft to get such a system were the F-15E Strike Eagle and the F-117A stealth fighter. There will also be the same kind of auto-stabilization system as on the Blackhawk, though with some major improvements.

- Electronics—The RAH-66 will introduce a new type of electronics packaging to Army systems. Instead of the "black boxes" that have characterized military electronics since the 1960s, all of the computers and electronics will be on sealed circuit "cards," slotted into several electronics bays. Each type of card is identical, with its specific function determined by the software that controls it. Thus, if an RAH-66 needs a computer upgrade, all the crew chief will need to do is install an additional computer "card" into the electronics bay. As an added benefit, other weapons systems will (theoretically) be able to use the same types of cards, thus simplifying maintenance and logistics requirements. All of the electronics systems on the Comanche are linked by a 1553 data bus, allowing one system to "talk" to another.

- Weapons—Pound for pound, the Comanche will be the most heavily armed aircraft in history. The basic weapon is a three-barreled 20mm Gatling-type gun (with 500 rounds of ammunition) in a nose turret. Along each side of the fuselage is a weapons bay with retractable door-mounts for the internal weapons. These can include Hellfire missiles, 2.75" Hydra-70 rockets, and air-to-air Stinger missiles. In addition a pair of stub wings, similar to the ESSS mounts on the UH-60 Blackhawk, can be installed to carry additional weapons and/or external fuel tanks. A normal combat load for an RAH-66 might be the 20mm gun, five AGM-114 Hellfire missiles, and two air-to-air Stinger missiles.

- Flight Controls/Maneuverability—The Comanche will be the first U.S. helicopter to make full use of a digital fly-by-wire control system like that of the F-16 Fighting Falcon. Also, the FANTAIL® rotor system allows it to turn faster than any other aircraft in the world. So versatile is the new system, that the FANTAIL® flying prototype (a modified Sikorsky S-76) achieved lateral (sideways) speeds of over 80 knots/130 kph!

- Maintenance—The measures taken to make the RAH-66 reliable and maintainable can only be described as fanatical. A built-in fault-isolation system automatically tells the crew chiefs exactly what is wrong (or right) with all of the Comanche's onboard systems. Maintenance has also been simplified, with an engine change taking only one hour. All field maintenance can be done with only six tools (carried in an onboard kit), while flight-line service requires only thirty-four.

- Deployability—With a pair of external fuel tanks, the Comanche will be capable of one-way hops of over 1,260 miles/2,286 kilometers. This means that it could self-deploy from the U.S. to anywhere in the world, albeit in a number of jumps. But more practical deployment options have also been built into the design, such as the ability to assemble or disassemble the main rotor (a five-bladed model, to keep noise down and efficiency up) and load or unload it aboard any one of a number of cargo planes in less than twenty-two minutes. The venerable C-130 Hercules, for instance, can carry one, while the C-5 Galaxy can carry up to eight, with the RAH-66s able to launch on mission shortly after arrival. In fact, the time required to service, refuel, and rearm a Comanche between missions is only fifteen minutes.

Putting all these pieces together will take some time, but when it hits the battlefields in the early 21st century, the Comanche will be the most dangerous rotorcraft in the world. Its stealth, armament and sensor package, and communications capabilities will probably make it the backbone of Army Aviation well into the middle of the 21st century. Keep an eye out for this new bird. I certainly will!

U.S. Army Personal/
Man-Portable Systems

One of the most coveted badges in the Army is a simple blue rectangle with an 18th-century Kentucky long rifle in silver. Called the Combat Infantry Badge, it tells you that you are looking at a soldier who has held a rifle in his hands, seen the face of battle, and fired at the enemy. A Navy can blockade an enemy, and an Air Force can demolish the enemy's economic and political centers, but ultimately it takes soldiers with their personal weapons to dig the enemy's soldiers out of their bunkers and trenches and send them packing. Individual soldiers and their personal equipment are the point of the spear for all the high technology that armed forces acquire and use.

A lot more is expected from the average soldier today than during the American Civil War, some 130 years ago. Back then, it was enough to be able to shoot another soldier with some accuracy at a range of perhaps 100 yards. Today's soldier faces threats that those boys could never have imagined on the killing fields of Gettysburg or Shiloh. The soldier on a modern battlefield (perhaps thousands of miles from home) may be assaulted by tanks, bombed by aircraft, or showered with toxic chemicals. In addition, that soldier must be clothed, fed, and told where to go and how to get there. And when he does get there, the bad news is that he may be asked to destroy a tank or shoot down an airplane or helicopter all by himself. The good news is that the Army and American industry have given him some of the best tools in history to do the job.

The Colt M16A2 Assault Rifle

The Marines have a saying that goes, "This is my rifle. There are many like it, but this one is mine." In that one statement is the embodiment of just what a personal weapon means to a soldier. It is their sword to slay their enemies—and ultimately, their reason for being. No other item of personal equipment is more essential to defining an individual as a soldier, not even the uniform.

The combat rifle used by modern soldiers is derived from two earlier weapons, the bolt-action rifle and the submachine gun. Both were used during World War II, though neither had all of the characteristics desired by infantry. Towards the end of the Second World War, the Germans developed the first assault rifle, the MP44. This was a fully automatic weapon, much like a submachine gun, but firing a more powerful, longer-ranged rifle-caliber bullet. In 1949, a Russian engineer named Mikhail Kalashnikov adapted the German design to produce the classic AK-47. The widespread adoption of automatic personal weapons, or assault rifles as they have become known, changed the nature of small-unit tactics. Instead of long-range marksmanship to wear down the enemy by gradual attrition, the goal became instant annihilation by concentrated and intense bursts of fire, with no need to aim so exactly. The vast increase in ammunition expenditure seemed cost-effective. Bullets are cheaper than soldiers.

The standard infantry weapon of the U.S. Army and Marine Corps today is the M16A2, a gas-operated 5.56mm automatic rifle that weighs almost nine pounds with a thirty-round magazine. If you look closely at a gas-operated rifle like the M16 or the AK-47, you notice a tube above the barrel that taps into the high-pressure hot gas that propels the bullet. When the rifle is fired, the gas is directed against a piston that works through a series of levers and springs to eject the empty cartridge case, cock the firing pin, load the next round, and close the breech. The design of the M16 is based on the Armalite AR-15, developed by Eugene Stoner in the 1950s. Stoner licensed the design to Colt Industries of Hartford, Connecticut, which initially produced it (as the CAR-15) for the U.S. Air Force Security Police in 1961. In 1966, the Department of Defense directed the U.S. Army to adopt it as a replacement for the 7.62mm M14 rifle. The M16 was four pounds lighter than the M14, and a soldier could carry three times as much ammunition. Though the 5.56mm bullet was smaller and lighter than the M14's 7.62, it had a much higher muzzle velocity, and it tended to tumble when it struck flesh. The theoretical range of the weapon is 550 meters, but most infantry combat takes place at much closer quarters, making elaborate scopes and sights a useless appendage (except for specialized precision sniper rifles, which require intensive marksmanship training to use effectively).

Made largely from stamped metal parts and plastic injection moldings, the M16 is relatively cheap to mass-produce. When it was first fielded in the 1960s, the troops nicknamed it "The Mattel Toy." And early on in Vietnam, it gained a bad reputation for jamming and misfires, mostly because the Army had substituted an inferior propellant in the ammunition, which caused excessive fouling. Reportedly, in one small Marine unit overrun by the Viet Cong near Khe Sanh in 1967, every man was found dead with a cleaning rod in his hand, trying to clear a stuck cartridge. Additionally, in the jungles of Southeast Asia, mud tended to get into the precision mechanism, making it difficult to close the bolt. To rectify these problems, a modified version, the M16A1, added a manual bolt-closure device, chromium plating on the chamber, and a slightly reduced rate of fire.

Combat experience with fully automatic weapons has shown that troops often hold down the trigger ("target fixation"), continuing to fire long bursts (the troops call it "rock and roll") after a target is hit or suppressed. At a rate of fire between 700 and 950 rounds per minute, a thirty-round magazine is emptied in less than three seconds. In 1982, the Army introduced the M16A2 as a solution. A selector lever lets the soldier fire single shots or three-round bursts. To fire another three-round burst, you have to pull the trigger again. This burst limiter saves vast quantities of ammunition without reducing the lethality of the

The M16 rifle can be fitted with an M203 grenade launcher, a single-shot, breech-loaded weapon that fires 40mm explosive projectiles. The grenade launcher has its own trigger and safety mechanism, and does not require any alteration of the M16. Typically, one soldier in each infantry squad is equipped with this "thump gun." *OFFICIAL U.S. ARMY PHOTO*

weapon. The M16A2 design introduced a selectable cartridge-case deflector, so that left-handed shooters would not be showered with hot brass. The M16A2 can additionally be fitted with an M203 40mm grenade launcher under the barrel for firing tear gas, smoke, or HE grenades. Usually, each infantry squad will have one M16 fitted with the M203.

A more unusual variant of the M16 is the M231 Firing Port Weapon. This is a short-barreled, fully automatic M16 without a front sight or fore-grip. The muzzle is designed to fit the firing ports of the Bradley Fighting Vehicle. In urban combat or in response to an ambush, spraying bullets from the sides of your Bradley might make tactical sense, but in practice, troopers leave the M231s stowed in the vehicle and carry standard M16s when they dismount.

Though the M16 is now entering its fourth decade of service and still going strong, there are plans for a new version, the M16A3. Budget constraints, however, will probably keep the -A2 variant in service well into the 21st century. There are many other 5.56mm assault rifles made by Heckler & Koch (Germany), Fabrique Nationale (Belgium), and even the Kalashnikov AKM (Russian design and manufacture, with copies built in China and many other countries). Each has its own relative advantages and disadvantages over the M16. But many of the soldiers I have talked to consider the M16 to be about the best compromise in 5.56mm combat rifles. Not perfect, but the best compromise. And in the near future, it will probably stay that way.

The M9 Beretta Model 92F 9mm Pistol

No weapon arouses more debate, or has a deeper emotional significance for its users, than the handgun. Though it causes very few casualties in combat, no other weapon (except perhaps the bayonet and the fist) is so up-close and personal. For most of this century, the standard handgun issued to members of the U.S. Armed forces was the Colt .45-caliber semi-automatic pistol. According to legend, this weapon was developed by the brilliant Utah inventor

The M9 9mm Beretta model 92F automatic pistol, which replaces the classic
.45 as the Army's standard pistol.
BERETTA USA

and gunsmith John Moses Browning (1855–1926), because the Army found that during the Philippine Insurrection (1900–1903), charging tribesmen struck by .30-caliber rifle fire would just keep on coming, oblivious to mortal wounds. The Army needed a weapon that would stop a berserk fanatic in his tracks. The .45 was heavy, mechanically complex, and had a massive recoil that made it hard to control, but no one ever complained about its lethality and stopping power. By the 1970s, however, the Army's stock of .45s was wearing out, and most European armies were adopting 9mm as the standard caliber for personal weapons. The 9mm bullet is lighter than the .45, but has a higher muzzle velocity. So at typical point-blank ranges, it is actually more lethal than a .45-caliber round (purists will always give you a lively debate on this point). It also has the advantage of being much more compact than a .45, so that a semiautomatic pistol magazine can carry more rounds.

After a long and controversial competition, the Army adopted the M9 Personal Defense Weapon, a 9mm semi-automatic pistol made by an American subsidiary of Italy's legendary Beretta, a family-owned firm that has been in the gun business for almost five hundred years. The M9 weighs 2.6 pounds, has a fifteen-round magazine, and has a better safety mechanism than the weapons it replaces. It was designed to be equally effective for left-handed or right-handed shooters. Reliability is awesome. In testing, three M9s fired 30,000 rounds without a jam or failure. So far, the biggest single problem with the M9 has been that the demand has exceeded the supply system's ability to fill the orders. The M9 is typically issued to soldiers who do not carry rifles, such as aviators and military police, though officers and special-operations forces are frequently assigned one for personal use as well. Generally, it is a nice piece to shoot, though the fans of the old .45-caliber will never give in to the folks they call "9mm Mafia."

The Javelin Anti-Tank Guided Missile

When portable anti-tank guided missiles (invented during the 1950s) first appeared in large numbers on the battlefield during the 1973 Arab-Israeli War, some "experts" predicted the imminent demise of the tank. Using the wire-guided Soviet AT-3 Sagger (the Russian nickname is Malyutka, meaning "little guy"), Egyptian and Syrian infantrymen knocked out hundreds of Israeli vehicles. Saggers came in a light metal suitcase with two missiles, a fold-out launch rack, and a periscope sight with a reusable control box. Despite its success, the Sagger had many flaws, as first-generation weapons frequently do. For example, the AT-3 was so slow (120 meters/sec) that an alert tank crew could see it coming. Also, it kicked up a big smoke and dust cloud on launch, and had to fly out at least 300 meters before the gunner could get it under control with his little joystick. This done, he had to steer it into the target: It might take half a minute to fly out to the maximum range of 1.8 miles/3 kilometers. As a result of these shortcomings, tankers quickly learned to keep a 360-degree watch on the horizon, and to take along an escort of mechanized infantry to provide suppressive machine-gun fire on anyone who popped up to fire a missile. While the effects of the Sagger were blunted by this change in tactics, a revolution in armored warfare had clearly taken place.

In the late 1960s, under the Medium Anti-Tank Weapon (MAW) program, the U.S. Army set out to develop a shoulder-fired wire-guided anti-tank guided missile as a long-ranged, highly accurate successor to the famous WWII bazooka. McDonnell Douglas won the MAW competition, and the missile was rushed into production as the M47 Dragon. This second-generation system was not a great success for several reasons. The gunner had to sit in an awkward position, hold his breath during the missile's twelve-second flight time, and try not to blink. The total weight of the missile, launcher, bipod, and control unit was over 30 lb/13.8 kg, the launcher kicked like a mule when fired, and the awkward bipod and firing position made it hard to track a moving target. Furthermore, its mighty back-blast made it almost impossible to fire from an enclosed space such as a bunker or cave. On the positive side, as long as the gunner kept the target centered in the crosshairs, the missile could reach out 1,200 yards/1,000 meters to accurately deliver its 5.4-lb/2.45-kg warhead (capable of penetrating 24 inches/610 millimeters of steel armor). On the whole, though, the Dragon's negatives outweighed its positives. Soldiers in the Army hated it, and the Marines preferred to fire their Dragons at enemy bunkers or buildings, which rarely move.

With the unpleasant experience of Dragon behind them, the Army Missile Command at Huntsville, Alabama, went back to the drawing board for a third-generation anti-tank missile. The winner of the competition for the new missile was the Texas Instruments/Martin Marietta Javelin. Javelin will be the first "fire and forget," shoulder-launched, anti-tank-guided missile to enter service anywhere in the world. With Javelin, there are no wires to litter the battlefield. The missile has an advanced "imaging infrared seeker" (an imaging computer chip similar to those used in video camcorders) that locks onto the target before firing, thus giving the gunner plenty of time to duck and cover before the doomed tank can fire back. On the downside, a complete Javelin system is heavy: almost fifty pounds, including the missile, the disposable launch tube, and a reusable day/night thermal-imaging sight/telescope/control unit. On the upside, though, the Javelin's range is around 1.2 miles/2 kilometers, and the gunner can select a direct flight path (at the center of the target) to hit bunkers and soft vehicles, or an arching "top attack" flight path to strike the thin roof armor of a tank. The warhead is a "tandem shaped charge," with an initial charge to strip away the outer layer of reactive armor (if present) and a main charge to attack the primary armor behind that. Performance figures for the warhead have not been released, but it is doubtful that any vehicle in the world can survive it. As an added bonus, since the Javelin's "soft launch" rocket motor reduces the recoil and back-blast at launch, the gunner can fire it from a standing, kneeling, sitting, or prone position. This also means that in urban or entrenched combat, Javelin can be fired safely from an enclosed space. Javelin is currently doing well in testing, and is expected to enter service with the Army and Marine Corps in 1995.

The Stinger Anti-Aircraft Missile

Ever since the first foot soldier was strafed by an airplane (probably somewhere on the Western Front in Europe around 1916), foot soldiers have dreamed of a weapon that would even the odds. If a plane flies low enough, and enough determined foot soldiers keep firing their rifles into the air, there is a small but finite chance that a bullet (pilots call them "golden BBs") will get lucky and hit a critical aircraft system or component, as many unlucky American aviators learned over Southeast Asia (1964–1973). Lucky or not, soldiers have always wanted a "magic" weapon that would let them reach out and sweep their aerial oppressors from the sky, and a man-portable SAM was just the thing. This idea seems to have occurred to Russian and American engineers at about the same time, in the 1950s, although the technology took years to develop. It required advances such as very sensitive infrared heat detectors, compact powerful rocket motors, precise miniature mechanical actuators for the steering fins, and finally, rugged miniaturized electronics to tie it all together.

The first man-portable SAM, the Soviet SA-7 Strela ("Arrow") missile, was introduced in 1966, and used (without much effect) in combat as early as 1967 by the Egyptians against the Israelis. The U.S. Army introduced its own man-portable SAM, the General Dynamics Redeye missile, in 1968. These first-generation weapons were "tail chasers" (meaning they utilized a lag-pursuit-intercept logic). The infrared seeker of the missile had to "see" the hot metal of the aircraft's engine exhaust, which usually meant that the target aircraft was already outbound, heading away from the missile shooter. If the target was flying faster than about 500 knots/800 kph, the missile would rarely be able to overtake it. Also, if the line of sight from the missile shooter to the target was too close to the sun, the missile might lock onto that very hot and quite unreachable star. Despite their limitations, these early man-portable SAMs did shoot down some aircraft, and thus had to be considered when plan-

A Stinger missile team of the 82nd Airborne Division in the Saudi desert. The goggles and cloth over the nose and mouth are standard desert gear.

ning air operations. Air forces faced with the threat of these heat-seeking missiles quickly developed flare dispensers and infrared jammers (like the ALQ-144 "disco ball"). If the pilot knew there were heat-seekers on his tail, he could drop a few flares, which, being a greater source of infrared energy than his jet exhaust, would deceive the incoming missile. But these were simple measures against a first-generation weapon, and the missile designers were already working on new "smarter" missiles like the Stinger.

The Hughes Stinger missile is a dramatic improvement in man-portable anti-aircraft weaponry. As a starter, Stinger's seeker head is cooled (using a compressed gas cartridge) before launch, to make it more sensitive to infrared radiation: It can see "hot" sections of the aircraft like the leading edges of the wings and glint off the canopy in addition to the engine exhaust. This means that the Stinger is an "all aspect" SAM, which can engage an airborne target from any direction: inbound, outbound, or crossing. The newest version, the Stinger-RMP, can even detect the "hole" an aircraft makes against the ultra-violet background radiation of the sky. Additionally, a combination of optical filters and computer signal processing gives it great ability to discriminate real targets from decoy flares. The maximum range may be as much as 11 miles/17 kilometers, depending on the speed and altitude of the target.

The missile itself weighs 23 lb/10 kg, and is about 60 in/152 cm long. A complete Stinger launch system weighs about 34.5 lb/15.7 kg, and includes the missile as well as a disposable launch tube that snaps onto a reusable "gripstock." The gripstock includes the aiming sight and launch electronics, with a socket for an expendable battery and coolant unit. An odd-looking boxy antenna is attached to the side of the launcher and plugged into a portable "IFF interrogator." This is basically a radio transmitter that sends out a series of coded digital pulses. If the target has a "friendly" IFF transponder that sends back the proper coded reply, you can assume it's friendly.

To use the Stinger, you start by removing a protective lens cap over the muzzle of the launch tube and inserting a fresh Battery/Coolant Unit (BCU) into the gripstock. This powers up the missile electronics and cools the seeker head. This done, you track the target through the telescopic sight. When the missile locks onto the target, you will hear a distinctive tone from a built-in speaker, and an indicator light appears in your sight picture. Once the seeker is locked on the target, you take a deep breath and pull the trigger. A small launch motor then ejects the missile from the tube to a safe distance, the guidance fins pop out, and the main rocket motor ignites. The missile accelerates rapidly up to over Mach 2 (1,300 knots/2,080 kph) and begins to intercept the target aircraft. When the Stinger gets to the target, the 6.6-lb/3-kg directional blast/fragmentation warhead (which has both proximity and impact fuses) detonates, spewing fragments towards the target. In the unlikely event that you miss, there is an electronic self-destruct mechanism, so that a live missile won't come down on friendly heads.

The Stinger is an excellent example of something that works and works well in the hands of a soldier. In Afghanistan, with minimal training and under very difficult conditions, the *mujahidin* guerrillas downed over 270 Soviet aircraft with Stingers, scoring an astounding success rate of 79%.

In addition to the shoulder-fired version used by the Army, Marines, Navy, and Air Force, Stinger can also be mounted on a number of helicopters and fighting vehicles. The twin launcher for helicopters weighs about 123 lb/55.9 kg, including the missiles, control electronics and coolant unit. It provides helicopters like the AH-64, OH-58D, and UH-60 and other Army aircraft with the ability to engage and kill enemy helicopters and fixed-wing aircraft in flight. Another Stinger carrier is the Avenger, developed by Boeing Aerospace and in production for the Army since 1990. The Avenger is a HMMWV with a compact turret mounting a pair of four-round Stinger launchers, a .50-caliber machine gun, and a digital fire-control system with a laser rangefinder and thermal viewer. It has the advantage of a cueing and data-link system. Finally, there is the new version of the M2 Bradley, the Bradley Stinger Vehicle, which is currently under test (as of spring 1994).

During much of the 1980s and early 1990s, the Army was almost completely dependent upon Stinger for tactical air defense of deployed ground units. Because of the cancellation of the DIVAD air-defense gun system and the gradual phase-out of older anti-aircraft systems (such as Chaparral, a ground-launched version of the famous Sidewinder air-to-air missile, and the 20mm M61 Vulcan cannon on an M113 chassis), the Stinger has been "the only game in town." With no anti-aircraft system between Stinger and the fixed-site Patriot missiles, it is fortunate that the U.S. Army has not had to face any enemy that presented a serious air threat. Until new systems designed to fill the gaps in the

Army's air-defense umbrella come on-line in a few years, tactical air defense will depend on Stinger, the little missile that can!

Clothing: BDUs, Helmets, Armor, and Chemical Suits

What does the well-dressed American soldier wear into battle these days? Well, though you might wonder at the fit of the clothes they wear, the troops the U.S. sends around the world are the best clothed in history. I'm not talking about the various dress or parade uniforms, but battlefield clothing—the stuff designed to survive in the dust of Saudi Arabia, the humidity of Panama, or the day-to-day grind in places like Germany or Korea.

At the moment, the basic field outfit is called the Battle Dress Uniform or BDU. It comes in a variety of different colors and patterns, as well as different weights depending on the climate. The basic varieties are:

- Forest—Used by most of the Army as their basic BDU. These use a forest/olive green as the basic color, with other darker colors in a pattern designed to break up the shape of a human against a background of woods or grass.
- Desert—This is the "chocolate chip" uniform made so famous by General Schwarzkopf during Desert Storm. These use a tan base fabric with various shades of brown and gray to help personnel "disappear" against the scrub brush and dunes found in the deserts of the world.
- Arctic—This is the new uniform for operations in mountain and cold weather environments. It is a combination of black, white, and grays that is extremely effective for hiding among the rocks and dirty snow piles so common on winter battlefields. It has an insulating lining and is somewhat heavier than the normal BDUs.

Soldiers applying camouflage face paint. Note the loose fit of the BDUs (Battle Dress Uniform) and the fabric cover over the "Fritz" Kevlar helmets. The small cuts and elastic band on the helmet cover are for insertion of leafy twigs and branches to provide additional camouflage.

OFFICIAL U.S. ARMY PHOTO

All of these uniforms come in a variety of sizes for men and women, and actually fit quite well, though they do tend to look rather baggy. The Army wants them to be comfortable, and not so confining that a soldier cannot jump or climb. In addition, all of the newest BDUs are treated with a waxy substance that resists absorbing chemical agents. There are also specialized coveralls for branches such as Armor and Aviation. Each of these uses Nomex® (a fire-resistant synthetic fiber from Dupont Corporation) as the primary fabric. Army coveralls seem to have dozens of pockets! And over the BDU goes the soldier's web gear, which is a general-purpose harness for ammunition pouches, first-aid kits, canteens, and the like.

Boots are almost as important as food to most soldiers. There are a number of different models. Of course, aviators and paratroops have their specific models, but it is the boots used by infantrymen that are so vital to the well-being of an army. Generally, the basic boot is comfortable and wears well enough. But the boots designed for extreme environments have traditionally given American soldiers aching feet or worse. Cold-weather boots have always been a particular problem, though the current version is adequate to the task. The new desert boot, introduced just prior to Desert Storm, proved to be a winner, and is quite popular with the troops in the field.

Since the human body is relatively fragile compared to bullets and shell fragments, the Army has developed some gear to help the exposed soldier survive. Helmets have been an important feature of the American soldier's outfit since World War I. Since that time, three different models have been worn. The original "flat dish" model, adopted from the British, was used until mid-1942. During WWII the Army adopted the classic "pot" shape that symbolized the GI until the early 1980s. Then the military broke with tradition and switched from metal to a synthetic called Kevlar (made by Dupont) as the basic helmet material. Kevlar is more than ten times as strong as steel by weight, and much easier to form into ballistically protective shapes. In addition, the Army did extensive research on the most effective shape for protective helmets, and the shape of the old German WWII helmets was found to be the best for preventing head injury in combat. Called the "Fritz" (an obvious reference to its German ancestry) by the troops who use it, it is the standard helmet issued today. The only improvement to the basic "Fritz" helmet is a new type of Kevlar (called Kevlar-29 by Dupont), which reduces the weight of the helmet a bit. This is important, as the weight of the "Fritz" can be hard on your neck muscles. The payoff, though, is the best helmet in the world, as over a decade of combat experience has proven.

A more recent development in protecting the American soldier has been the adoption of body armor—or "flak jackets," as they are known. First used in Vietnam, they greatly reduce the likelihood of fatal wounds to the chest and torso. Basically, this is a vest with insertable panels of Kevlar; and it looks very much like a down jacket. Early models were heavy, stiff, and confining to the wearer, but they radically reduced fatalities in units where the soldiers wore them regularly. The newer vests are much lighter and more flexible, though still a bit binding. Nevertheless, all the soldiers I know wear theirs religiously—to avoid what they call "unnecessary perforation"!

It is strange that today—almost five years after the dismantling of the Berlin Wall and the end of the Cold War—U.S. soldiers have probably never been under a greater danger of attack by chemical weapons. Though the dictators of the world are doing their best to acquire nuclear weapons (the number-one status symbol among dictators), several of them have settled on an older, and somewhat lower-status technology—chemical and biological weapons—as a sort of "poor man's atomic bomb." Because Saddam Hussein had actually used his chemical weapons against the Iranians and the Kurdish people, when American forces deployed to the Persian Gulf in 1990, they took the threat of attack by Iraqi chemical weapons extremely seriously.

Luckily, so had the Department of Defense. In addition to the acquisition of the Fox vehicles from Germany, they had recently introduced a new chemical protective garment for use by U.S. forces. Previously, the standard U.S. chemical suit (more of a rubberized smock actually) was bulky, uncomfortable, and almost impossible to work in for anything more than an hour or two without suffering heat prostration.

The new suit was a major improvement over the earlier model, resembling a quilted jumpsuit with thick insulation. The outside is covered in a camouflage cloth similar to that of the BDU, and together with the layered inner protection of the suit, it is resistant to virtually all known chemical agents. On the inside is a fabric shell containing an activated charcoal liner to absorb moisture and keep the occupant (relatively) dry, and consequently (relatively) cool. In fact, soldiers who wore them during Desert Storm actually found them quite comfortable, though some of this was due to the fact that Desert Saber, the one-hundred-hour ground war, was fought during a period of cold and rain! The downside of these new suits was that after being worn for something like five continuous days, the inner shell and charcoal liner literally peeled off onto the skin of the wearer. The newer models have been designed to avoid this problem, and seem to work well.

Soldiers on operations wear the suits whenever there is any danger of chemical attack. In the event of a suspected chemical attack, each soldier immediately dons a protective mask and hood, as well as a set of booties designed to keep the soldier from fouling—or "sliming"—his boots with toxic agents. The mask is an improved design with replaceable filters (they have an easier "draw," making it less fatiguing to breathe), and even a small port to take water from a standard canteen. The suit is by no means comfortable, but it's a vast improvement over past designs, and a much better alternative to becoming a chemical toxin casualty.

So when will General Sullivan give every soldier a GPS receiver and a personal IVIS terminal? (Remember the Space Marines in *Aliens*?) Well, don't laugh yet, because such things are on the way, and will probably be part of the soldier's outfit before the end of the first decade of the 21st century. For example, Trimble Navigation, of Sunnyvale, California, is already working on integrating a miniature GPS antenna and receiver into a standard "Fritz" helmet. And biomedical monitoring and transponder gear will probably be part of a soldier's standard kit by 2010. If this sounds far-fetched, stay tuned: Things like fully powered body armor with air-conditioning and environmental controls are already being discussed at places like Fort Benning, Georgia (the Army Infantry Center), and Fort Detrich, Maryland (where they work on habitability and sustenance technology). There *will* be Starship Troopers!

Food: T-Rations and MREs

"An army travels on its stomach." —Napoleon Bonaparte

Nothing in soldiers' lives so affects their well-being and morale as the quality and quantity of their field rations. The problem with food is that being organic, it tends to spoil. So ever since people first left their caves to make war on their distant neighbors, warriors have been trying to find better ways to package food and make it last. In the time of Alexander the Great, goatskins and ceramic vessels provided the packaging for food and drink, albeit not terribly effectively. Napoleon awarded prizes to inventors who demonstrated the ability to preserve foods in glass jars. But the basic packaging technology that would be the standard for the next 150 years, the tin can, was a British invention—with everything from canned butter to boned pheasant being canned and shipped to British troops around the world. However, in those days it was expensive to package food in this way; and only delicacies and supplements like butter and condensed milk ever found their way to the forward troops. The desire to keep down costs and maximize the efficiency of the ration system perhaps hit a low point during the American Civil War. The malnutrition that resulted from the basic diet of salt pork (bacon), hardtack (unleavened bread), and black coffee may have caused as many deaths as bullets did. Only the desperate intervention of citizens' organizations like the United States Sanitary Commission and the Red Cross avoided even more grievous losses.

Over the next century and a half, advances in canning technology and reduced costs of canning (due to civilian consumption) contributed to better quality and longer shelf life. And in due course, the U.S. Army attempted to produce nutritious and tasty canned and packaged rations. The best known of these, the famous C and K rations of World War II, provided individual portions of well-known foods in cans and sealed packages, with shelf lives measured in years. Some, like the canned beef stew and spaghetti, were almost identical to civilian products from Armour or Campbell's. Others, like the hideous pork patty (sausage in grease gravy), and the notorious fruitcake (a sort of sweet and sticky hockey puck), were usually passed along to captured enemy troops. Because of their mixed reception by front-line combat troops, the Army felt obliged to supplement or replace the field rations with fresh food as often as possible. This policy—admirable in principle—could nevertheless be taken to silly excesses, as in Vietnam, when jungle patrols might have chilled cases of beer airdropped to them, and remote firebases greeted returning patrols with full-course steak dinners and well-stocked bars. Clearly, in the attempt to make field rations as palatable as possible, the Army supply system had gone berserk and needed to be fixed. Luckily, a new packaging technology had come on the scene to help the military with this problem.

The coming of the space age and man's early ventures into orbit and to the moon meant that he had to take food and drink with him. At first this took the form of pureed foods in toothpaste tubes and crumbly crackers and cookies. But the astronauts' distaste for such stuff, and the high public visibility of the space program, forced NASA to research and develop better food products to keep flight crews happy and healthy. At first, they tried freeze-dried foods—quick-frozen and then placed in a vacuum to remove all the moisture. But this did not work well with meats and baked goods. And by the end of the Apollo moon-landing program, NASA was allowing common grocery items like bread slices, canned meats, and peanut butter and jellies on lunar missions. The real breakthrough came with the development of the "wet pack," a sealed plastic bag with dehydrated foods such as meat slices, stews, or vegetables, sterilized to prevent spoilage (usually by steam heating or radiation bombardment), and then rehydrated for use by the crews. The same techniques were applied to other types of prepared foods (lasagna, chicken and rice, etc.) for larger institutional-sized containers. And thus the technology behind the T-ration and the MRE was born.

The current U.S. Army strategy of feeding soldiers in the field is based around the following three types of rations:

- A-Rations—These are fresh foods, procured locally from around the operational area, and prepared by standard Army field kitchens. This is the cheapest and most preferred type of ration (by both the soldiers and the Army), though local vendors and supplies may limit availability.
- T-Rations—These are prepared foods, from vendors like Stouffers and Swanson, packaged in large aluminum trays, matched into meal sets for groups of twelve soldiers, and then heated in buffet-type heaters with boiling water. They usually do not require refrigeration, though some special meals (like the famous 1990 Thanksgiving dinner during Desert Shield) may require cold storage in transit.
- Meals Ready to Eat (MREs)—This is the standard field/combat ration of the American military. An MRE is a series of wet, dry, and freeze-dried food packs, with an accessory package (spices, a spoon, fork, napkins, etc.) sealed inside a rugged (some say too rugged!) brown plastic bag. There are twelve basic varieties, with one of each variety being packed in a case of MREs. Each MRE contains about 3,000 usable calories of food, and each

The contents of a Meal Ready to Eat (MRE). This particular one has pouches of chicken with rice, cheese spread, crackers, drink mix, coffee, and a cookie bar. Note the small bottle of Tabasco sauce, a favorite of the troops. *JOHN D. GRESHAM*

soldier in the field is allocated four per day under the current Army supply scheme.

The T-rations have been a godsend to the Army in terms of cost and portability. When you open up a T-ration box, you usually find three aluminum trays of food—a meat entree, a starch dish, and a vegetable dish. Also in the T-ration box is that most prized of Army foodstuffs, an oversized bottle of McIlhenny Co. Tabasco sauce! In all deference to the Army and their contractors, the foodstuffs supplied to the troops can be a bit on the bland side, especially to some ethnic groups that have found a career in the Army so attractive. So the addition of spicy condiments has become an essential part of the T-ration specification. Overall, the T-ration program has been a success, though everyone has his own preferences. For example, one young cavalry officer that we spoke to had nothing good to say about the chicken breasts in honey glaze sauce over rice. But a senior supply officer of a cavalry regiment declared the dish "delicious," and said that he could eat it every day! Like the population that makes it up, the Army is a mixture of tastes.

Because of the Gulf War, MREs got a very bad rep. Called at times "Meals Rejected by the Enemy," they garnered many negative comments during the Gulf crisis. Part of the reason for this was the limited variety available to the troops during the early days of Desert Shield. During the dark days of August 1990, before the Army logistics and support services caught up with them, the first troops deployed to the Persian Gulf (mainly the 82nd Airborne and 101st Air Assault Divisions) had nothing to eat but MREs. At that time, there were only four varieties of MRE (as opposed to the dozen of today), a choice that was made worse by the dietary requirements of our Saudi Arabian allies. Prior to Desert Shield, the Saudi National Guard was primarily a security force charged with the protection of mosques around Mecca and other holy places. They were not a field army, and so lacked the support structure to maintain themselves in the desert of northern Saudi

Arabia. The Saudis did not have a stockpile of ready-to-eat field rations for their ground troops, and lacked field kitchens to cook food for them. So the Saudis asked if they might buy several million MREs as interim field rations, pending delivery of their own field kitchens. This was cheerfully done. After the MREs were delivered, though, someone (nobody seems to know whether it was Saudi or American) realized that two of the four MRE types contained pork (ham in one, roasted pork in the other), and thus were forbidden for Muslims. To avoid embarrassing us, the Saudis kept the two varieties that did *not* have pork in them, and graciously donated the rest to the troops of the XVIII Airborne Corps already on the line up on the Iraqi border. Thus it came to pass that the soldiers of the XVIII Airborne Corps wound up eating ham and egg omelet and pork patty MREs for some weeks. They were positively sick of them! Nevertheless, everyone did eat, and they got on with the business at hand, the winning back of Kuwait.

From this slight dietary debacle came the initiative to greatly expand the variety and quality of the MRE program. The first move was to develop and package a number of new types of MRE. The next step, which is still going on, was to look beyond the traditional foods that have been packaged into field rations and produce MREs that better reflect the eating habits and tastes of the young Americans that make up the raw material of the U.S. Army. More about the new MRE technology later, but first let's look at the existing variety of MREs available for use by the American soldier.

If you were to open a case of MREs—and there is only one kind of MRE package in late 1993—you would find one of each kind of MRE inside. This was done so that nobody could complain that the Army and its contractors were trying to force MREs of one kind or another onto the troops. So, in what is probably the first rule of MRE consumption etiquette, when the troops are issued their MREs, they randomly reach into the box, and just pull one out. In this way, no individual can claim he was "gypped." The second rule is that it's OK to "trade" for another MRE (like "brown bag" rules in grammar school). Once you've got your MRE, opening the package (each weighs about 2 lb/1 kg) takes persistence; the brown plastic of the bags is so tough it's almost bullet-proof. As a result, the soldiers tell me that a Swiss Army knife (with a built-in scissors) is considered an essential tool for the MRE gourmet. Now, in case you wonder just what is inside these little brown plastic packages, consider some of the following menus:

- Menu No. 2—Corned beef hash, freeze-dried pears, crackers, apple jelly, oatmeal cookie bar, beverage base powder (fruit drink), cocoa beverage powder, Accessory Package "C" (Taster's Choice coffee, creamer, sugar, salt, pepper, gum, matches, hand cleaner, and toilet tissue), and a spoon.
- Menu No. 4—Omelet with ham, potatoes au gratin, crackers, cheese spread, oatmeal cookie bar, beverage base powder (fruit drink), Accessory Package "C" (Taster's Choice coffee, creamer,

sugar, salt, pepper, gum, matches, hand cleaner, and toilet tissue), Tabasco sauce, and a spoon.

- Menu No. 7—Beef stew, crackers, peanut butter, cherry nut cake, Accessory Package "A" (Taster's Choice coffee, creamer, sugar, salt, pepper, gum, matches, hand cleaner, and toilet tissue), Tabasco sauce, and a spoon.
- Menu No. 8—Ham slice with natural juices, potatoes au gratin, crackers, apple jelly, chocolate-covered brownie, beverage base powder (fruit drink), cocoa beverage powder, Accessory Package "A" (Taster's Choice coffee, creamer, sugar, salt, pepper, gum, matches, hand cleaner, and toilet tissue), Tabasco sauce, and a spoon.
- Menu No. 11—Chicken and rice, crackers, cheese spread, chocolate-covered cookie bar, beverage base powder (fruit drink), Starburst candies, Accessory Package "A" (Taster's Choice coffee, creamer, sugar, salt, pepper, gum, matches, hand cleaner, and toilet tissue), Tabasco sauce, and a spoon.

Once the package is opened, you collect your beverage (usually water, heated for coffee or chilled for the drink powder, or boxed milk). Then it is a matter of organizing what you have found in the bag. Should you desire to heat your entree package, you can boil it in water (if available). But true MRE gourmets tell me there is a much preferred method. Step one is to find a friendly HET driver and drop the MRE entree package down the exhaust stack while the engine idles. You wait for ten minutes (timing here is essential), then have the driver gun the engine. The MRE package blows right up out of the stack, perfectly heated through! Another option, though one shunned by many troops, is the use of a little Army issued gadget, the U.S. 1992 MRE Heater. This is a catalytic mitt, which, when activated with water, produces enough heat to make an MRE entree pack warm enough to enjoy. These are also used a lot in Arctic regions just to thaw frozen MRE packs. The downside is that the heaters produce hydrogen (an explosive gas) as a by-product of the catalytic reaction (this means no smoking or open flames around them), and the other by-products of the reaction are somewhat toxic and have to be disposed of carefully.

So, you might ask, what are they like to eat? Not bad actually. You have to eat the entrees out of the wet packs, which tends to be a bit messy (a tip—slit the bags the long way to reduce the mess), though quite practical.

In general, MREs are tough for the supply system to handle. Their contents are hydrated with so much water, they are relatively heavy and bulky. And of course, they make a lot of waste. In view of all the garbage generated by the MREs (the Army calls it "wet" trash or garbage), it is a good idea to use the brown plastic MRE bag to stuff all the waste in. This wet waste is a major problem in using MREs, as the current environmental policy of the Army is to treat the lands that our forces enter at least as well as our own domestic exercise areas. This means that trash must be packed out or buried in an approved waste site.

Yet for all their problems, until the Army figures a way to make water out of thin air in the desert, MREs will remain the best compromise available. Because they require very little water to supplement the meal, the MRE is going to be field ration of choice for U.S. military forces when they operate away from home.

Which brings us to our discussion of future MRE developments. One of the main goals of the Army is to make field rations that are both attractive to the soldiers and nutritious. That second goal has been achieved quite well with the current MREs. Well balanced nutritionally, particularly in mineral content (which is vital in areas where soldiers sweat a lot), each MRE delivers about 3,000 usable calories (if fully eaten), with four MREs being allocated for each soldier each day. Surprisingly, because of the high caloric content in each meal, troops in the field actually have the tendency to gain weight (which is almost unheard of in military history), despite the heavy workloads placed upon them in field operations. But this does not solve the problem of variety and taste. In addition, with the growing diversity of personnel in the Army (Muslims are the fastest-growing ethnic group in America, and in the Army), it is becoming necessary to produce field rations that meet the strict dietetic requirements of groups like vegetarians and Muslims. In late 1993, a new series of MREs—with entrees based around vegetable products like lentils and potatoes—was produced and airdropped as relief supplies for the Muslims in Bosnia. As for more common American tastes, there are some promising efforts to package "fast" foods like hamburgers, as well as Mexican and Chinese entrees. But the crown jewel of these efforts is (yes, you guessed it!) an MRE with a slice of pizza in it. A specially shaped thermal mitt heats the slice in its wet pack and melts the cheese. These new "fast food" MREs will probably be getting to the troops in the field in the next few years, and should be quite a hit. Nevertheless, one old cavalry sergeant that we ate with at Fort Bliss was heard to say that the pizza MRE would not be complete until the contractor found a way to put a self-chilling can of beer inside! One thing that is certain, though. The Army is spending a lot of money to make sure that the U.S. Army is the best fed in the world. *Bon appétit!*

Radios: The SINCGARS Family

In May of 1940, when the Nazi Panzer divisions invaded France, most of their tanks were inferior in firepower and protection to the French and British tanks opposing them. But *every* German tank had a radio, while only special Allied command vehicles were so equipped. In maneuver warfare, every combat unit must be able to do three things: move, shoot, and *communicate.* The German Army gained tremendous tactical and operational advantages in flexibility and in command and control from those fragile, short-ranged vacuum-tube sets. Other armies paid careful attention to this lesson. They still do.

The backpack version of the SINC-GARS jam-resistant frequency-hopping radio. Dismounted cavalry scouts might carry this type of radio.

Revolutionary advances in electronics during the 1970s and '80s left the U.S. Army with a collection of obsolete radio sets that were heavy, and hard to maintain, drew too much power, and put out too much heat. They were often incompatible with frequency bands and transmission modes used by the Navy and Air Force. Even worse, they were vulnerable to enemy interception and jamming. The Russians, with decades of experience in jamming Western radio broadcasts (Voice of America, Radio Free Europe, etc.), had made "radio-electronic combat" a key feature of their tactical doctrine.

In the early 1980s the U.S. Army specified the design for a new SINgle-Channel Ground and Airborne Radio System (SINCGARS). SINCGARS (which entered service in 1988) is a family of compact, lightweight, reliable, and secure FM radios that can use any of 2,320 different frequencies between 30 and 87.975 MHz in the VHF band. The Army plans to procure 150,000 SINC-GARS radios from General Dynamics (San Diego, California) and ITT Aerospace (Fort Wayne, Indiana), with additional orders from the Marine Corps and various government agencies. The system resists jamming by using "frequency hopping": The transmitter and receiver are synchronized to jump between widely spaced frequencies at very short intervals. To defeat the system, the enemy would have to radiate a tremendous amount of energy spread across a large slice of the electromagnetic spectrum. There are eight basic models:

- AN/PRC-119—This is a backpack model, capable of being carried by a man.
- AN/VRC-87—A vehicle-mounted, short-range model.

- AN/VRC-88—A vehicle-mounted, short-range model that can be dismounted if desired by the crew.
- AN/VRC-89—A vehicle-mounted, long- and short-range transceiver model.
- AN/VRC-90—A vehicle-mounted, long-range transceiver model.
- AN/VRC-91—A vehicle-mounted, long-range model that has the option of being a dismountable short-range transceiver.
- AN/VRC-92—A vehicle-mounted, dual-channel (essentially two radios together), long-range model with retransmission (over separate radio net) capabilities.
- AN/ARC-201—The standard helicopter/aircraft transceiver model.

The AN/PRC-119 backpack portable model weighs about 22 lb/10 kg. The short-range models have a power output of about 5 watts, and a maximum range of between 2.5 to 5 miles/4 to 8 km. Long-range models have a power output up to 50 watts, and a maximum range of 5 to 22 miles/8 to 35 km. These relatively low-power output levels make it harder for an enemy to detect and locate the transmitter. All of the SINCGARS systems can handle voice, text, digital, and data communications (just hook up the desired transmission device), and even fax transmissions with the appropriate attachments. The latest models also have built-in cryptographic (scrambling) units for added security. During Desert Shield and Desert Storm, SINCGARS stood up to heat, dust, sandstorms, and bad weather with exceptional reliability. They should be the standard Army communication package well into the 21st century.

The NAVSTAR GPS System

To wage a successful war of maneuver, a commander must constantly know the answer to two questions:

Where the hell is the enemy?

Where the hell am I?

Military history is filled with defeats that resulted after bold, aggressive flanking columns got lost in the woods, or stouthearted defenders dug in on the wrong hill. Ever since reasonably accurate terrain maps were first drawn up (around the beginning of the 18th century), armies have tried to teach junior officers the art of map reading and navigation. The results have often been disappointing. The advent of modern electronics has provided some limited advances with gyrocompasses and inertial navigation systems, though their high cost limited their usage. Early satellite navigation systems also had promise, though their cost, and the cost and size of receiver sets, made them unusable by most of the military. Something new was needed to provide exact position accuracy. That something was the NAVSTAR GPS system.

The NAVSTAR Global Positioning System (GPS) is a dramatic advance in navigation. It starts with an array of twenty-four satellites (called a constellation) 10,900 miles/17,600 kilometers up in an orbit inclined at an angle of 55° to the equator. At this altitude, it takes a satellite twelve hours to circle the earth. When all twenty-one primary satellites (plus three spares) have been launched, at least four will always be visible to a receiver virtually anywhere in the world. Each satellite carries a super-accurate "atomic clock" and a low-powered transmitter that broadcasts specially coded time signals and status messages on radio frequencies of 1227.6 and 1575.42 MHz. By correlating the signals from at least four satellites, and doing some fancy trigonometry, the computer in a small portable receiver can determine your location, altitude, speed, and time with great precision. Relatively inexpensive civilian GPS receivers are typically accurate to within 25 meters/82 feet of 3-D positional accuracy. And military receivers, which can decode a more accurate encrypted part of the GPS signal, called the P(Y) code, get much better performance. The original GPS specification demanded accuracy with a P(Y)-code receiver to 16 meters/52.5 feet, though positional accuracies of around 5 meters/16.4 feet are considered typical by military GPS users. As an aside, by linking several GPS receivers and a radio transmitter at a known (i.e., surveyed) geographic point, surveyors can determine positional accuracy down to plus or minus one centimeter/.4 inch! As an added bonus, since every GPS receiver automatically synchronizes itself to the super-accurate clocks on the satellites, combat units can now precisely coordinate their actions in time as well as space. Handheld military GPS receivers as small as 14 oz/397 grams are already available, with the ability to display information (in six languages!) about the exact position of the sun and phase of the moon on any given day.

Now you might ask, since GPS receivers are commercially available from many American, European, and Asian electronics firms, what prevents an enemy from buying and using off-the-shelf units to gain the same kinds of tactical advantages? The GPS system was designed to provide "selective availability" during a crisis or conflict. When the GPS satellites receive a special encoded command from Air Force ground controllers, they can start broadcasting less accurate data. Then, unless you have a military P(Y)-code receiver, and the proper cryptographic key for the day, you

The Small Lightweight GPS Receiver, which enabled the Army to navigate the deserts of Iraq. The square bulge on the top of the case is the antenna, which can detect transmissions from five NAVSTAR satellites simultaneously. To receive the highly accurate P (Y) code, the device must be loaded with a "crypto key." The command "Zeroize Cryptokeys" erases this secret information. *Trimble Navigation*

will only be able to determine your position within about 100 meters, rather than 25 meters. In fact, it is possible for the controllers to selectively degrade (say, to only a 1,000-meter/.61-mile accuracy) the non-P(Y) accuracy of the system on a local basis, such as over the Middle East during a time of conflict, if that is desired.

So how does one make use of this amazing little bit of technology? Well, consider the following. The AN/PSN-10 (V) TRIMpack GPS receiver is the best-selling military GPS receiver in service today. Literally thousands of them were used in the Persian Gulf during Desert Shield and Desert Storm, with more than 18,000 units in service worldwide since then. A civilian version, called the Scout-M, is available for anyone who cares to use one for their 4WD or bass boat. Made by Trimble Navigation, the Centurion, the newest P(Y)-capable version, weighs 3.1 lb/1.4 kg, and is about the same size as a good pair of binoculars. The flat antenna is built into a nearly indestructible green plastic case. It has a removable power pack with NiCad rechargeable batteries (non-rechargeable Lithium batteries were used in earlier units). On the front panel there is a backlighted (suitable for use with the new AN/PVS-7 low-light goggles) four-line LCD display panel, a rotary selector switch for different operating modes, and two toggle switches for changing the various options on the display.

The unit also has a serial data port that allows it to communicate with any compatible computer, digital system on a vehicle or aircraft, or even other GPS receivers. During Desert Shield the Army acquired 8,000 of these Small Lightweight GPS Receivers (SLGR—the troops call them "sluggers" for short) for about $3,600 each under an emergency procurement. The SLGR can be programmed with the locations of up to 1,089 "waypoints"; and by simply flipping the knob to R+A (Range and Azimuth), you can read out your current range and bearing from any three waypoints (such as an enemy position, a friendly base, or a logistics base). During Desert Shield, SLGRs were initially so scarce, and so vitally needed in the trackless desert, that many personnel tried to buy them directly from the manufacturer, using their own credit cards. Special Forces teams that operated deep inside Iraq, and a few pilots shot down in enemy territory, credited their survival to the precision of their GPS receivers, which enabled them to link up with friendly helicopters in exactly the right place at exactly the right time.

Using the SLGR is simple in the extreme (please note that I was using a non-P(Y)-code SLGR for the following) so much so that during Desert Shield, soldiers who got them just ripped open the packing containers, and were using them within half an hour. You start by turning the selector knob from the OFF position to Status and Setup (STS). As soon as the startup screen clears, the STS readout screen indicates the following data:

```
Tracking 0 SVs
GPS n/a
```

```
Battery used: 0:00
INT antenna <more>
```

This means that the SLGR is not yet tracking any satellite vehicles (SVs). Thus, there is not a GPS position yet available, as shown by the GPS n/a indicator. The Battery used indicator tells us how much time has been logged since the last charging; and the INT antenna indicator shows that we are making use of the internal antenna, as opposed to an externally mounted one. After a minute or two you will begin to see the SVs counter begin to change. Once it reads:

```
Tracking 4 SVs
GPS OK
Battery used: 0:03
INT antenna <more>
```

you have a three-dimensional fix (two-dimensional fixes are possible with just three satellites) and can start to work. One of the first things that you might want to do is set up your preferences for using the SLGR. By clicking the horizontal L/R toggle switch, the <more> indicator will begin to blink. Now, by clicking the vertical INC/DEC toggle switch down twice, you get to the settings screen. You will probably see something like this:

```
Datum: WGS-84
Time: UTC
Units: Metric/DEGS
Mode: DMS/Tr <more>
```

The Datum: WGS-84 indicator tells you that the SLGR is currently using the WGS-84 Merchich mathematical model of the earth to figure its coordinate readouts. Different maps use different earth models as their point of reference. Suppose that we wanted to take a walking tour of Washington, D.C. If we acquire a 1:50,000 tactical map (Sheet 5561 I, Series V734, Edition 1-DMA, Alexandria) of the area from the Defense Mapping Agency (they are available through the National Geological Survey and NOAA), and we look at the legend of the map, we find that it conforms to the 1927 North American Datum. So, we click the L/R switch once, and the Datum indicator begins to blink. Now, using the vertical switch, we scroll through until we get to NAD-27, CONUS. Clicking the L/R switch twice now, we get to the Time: setting. This allows you to select either UTC (Universal Time Code—Greenwich Mean Time) or Local (your current time zone) as your clock readout. Leave them at UTC for the moment. Clicking the L/R switch again, we can select the read-out units, in this case ENGLISH/DEGS (English and degree units). This done, we again use the L/R switch to set the Mode indicator to Degrees-Minutes-Seconds (other options include UTM and the military grid reference system MGRS, as well) and True North, so that it reads DMS/Tr. This done, you can now use the L/R and INC/DEC switches to get back to the basic STS screen. The settings that you have just made will now be the defaults for the system until you again change them.

Now, if you turn the rotary switch to POS, the receiver will display a time and position. A fix on the front steps of Union Station (I love to travel on trains!) on a Saturday afternoon (EST) might show as follows:

```
SAT UTC16:00:00
lat 38°53'56.7"N
lon 77°00'23.6"W
alt + 260ft ±300
```

A quick review of the map shows that to be right in front of the station, just south of the Metro station. If we were to go back to the STS setting and reset to the Military Grid Reference System (MGR) setting, then go back to the POS setting with the selector knob, then the readout would look something like this:

```
SAT UTC16:00:00
MGRS 18S
  UJ 25988 07562
alt + 260ft ±300
```

If you had one of the maps drawn to this system, specifically sheet 18S, you would be able to see down to the square meter where the unit was telling you your position was. But let us assume that we want to use the GPS receiver to set up a tour for others to follow. You now select the Waypoint (WPT) setting for the selector knob, and you will see the following display:

```
SAT UTC16:00:00
lat 38°53'56.7"N
lon 77°00'23.6"W
alt + 260ft <fix>
```

By clicking the L/R switch to the left, you can make the <fix> indicator blink. Once you have done this, a downward click of the INC/DEC switch will automatically insert this position into the waypoint array of the SLGR as waypoint AA (the first of 1,089 possibles) like this:

```
WPT AA ‡ 160000
lat 38°53'56.7"N
lon 77°00'23.6"W
alt + 260ft <fix>
```

This done, we walk southwest across Constitution Avenue to the north steps of the National Air and Space Museum (one of my favorite places on earth!) on the north side of Independence Avenue. Taking another fix, we find the POS readout to be:

```
SAT UTC16:30:00
lat 38°53'25.0"N
lon 77°01'06.2"W
alt + 90ft ±300
```

Another check of the map shows the receiver to be generating good fixes. We add the new position to the waypoint array, and the readout shows:

```
WPT AB ‡ 163000
lat 38°53'25.0"N
lon 77°01'06.2"W
alt + 90ft <fix>
```

Our next waypoint fix point is the west steps of the Capitol building:

```
WPT AC ‡ 164500
lat 38°53'19.1"N
lon 77°00'47.6"W
alt + 9ft <fix>
```

Note the sudden drop in altitude, though it still is within the accuracy tolerances (100 meters/328 feet). We move on and add waypoint fixes for the Washington Monument (east parking lot):

```
WPT AD ‡ 171500
lat 38°53'25.0"N
lon 77°01'06.2"W
alt + 90ft <fix>
```

Moving west again, we walk up the reflecting pool, past the Vietnam Veterans Memorial (be sure to stop and see the black wall), and up to the steps of the Lincoln Memorial to get our next fix:

```
WPT AE ‡  174500
lat 38°53'16.3"N
lon 77°02'56.7"W
alt + 497ft <fix>
```

Continuing west again, we walk across the Memorial Bridge over
the Potomac River to the entrance to Arlington Cemetery and our
final waypoint:

```
WPT AF ‡  181500
lat 38°53'03.4"N
lon 77°03'57.2"W
alt + 238ft <fix>
```

With the six waypoints now stored, it is possible to use the stored positions to get actual guidance information for our return walk to Union
Station. For example, if we switch to the R + A setting, and use the L/R
and INC/DEC settings to display the data for waypoints AA (Union
Station), AB (the Air and Space Museum), and AC (The U.S. Capitol), it
should look something like this:

```
R+A: AA  AB  AC
azm 084° 088° 078°Tr
rng 3.5 2.4 2.9 Mi
vrt 22 -148 -220 Ft
```

This shows us that we are 3.5 miles from our starting point (based on a
great circle navigation plot—essentially as the crow flies) at a heading of 84°
true (as opposed to magnetic). Now suppose that you want to use the SLGR to
dynamically navigate you back to Union Station, on-the-fly as it were. To do
this, turn the knob to the NAV mode setting. You should see the following:

```
TO: AA ttg *
vel 0MPH 084° Tr
rng 3.5Mi azm084°Tr
vrt 22Ft <more>
```

This tells us that to navigate to waypoint AA (Union Station) we need to
move along a heading of 084° true (roughly east along the mall). The velocity
(vel) readout of 0MPH tells us that we are not moving yet, and the time to go
(ttg) readout of * means that we have not yet started our trip home to the train

station. As soon as we begin to walk (most humans walk at about 5 mph/8.2 kph), the SLGR will begin to calculate `vel` and `ttg` numbers for the readout (anything over 3 mph/5 kph will give the SLGR the necessary Doppler to begin calculating these readings). Almost immediately, the following readout should appear:

```
TO: AA ttg 42:00:00
vel 5MPH 084° Tr
rng 3.5Mi azm084° Tr
vrt 22Ft <more>
```

This tells us that if we maintain a straight line to the train station (unlikely in Washington traffic, but what the hell), and walk at our current speed all the way, we can get back in forty-two minutes. Now, the SLGR will continuously update these figures, and if we stop and look at something else, will continue to guide us home. Later, if we want to add other known places for waypoints, we can insert them via the toggle switches on the front panel of the SLGR. In addition, we can connect our SLGR to another, and dump the waypoints to anyone else with one of these clever little devices.

You might ask, what does all this have to do with tactical operations in the field? More than you might think actually. Consider the following story. Prior to the beginning of Desert Storm, special forces teams from the United States, Great Britain, and other Coalition Allies went into Kuwait and Iraq armed with their usual array of weapons, as well as some of the little SLGRs doing exactly what we have been doing, taking readings and fixing waypoints. These found their way to the SLGRs of the cavalry officers in 2nd and 3rd ACRs, so that they could program their own SLGRs to guide them to the phase lines and road junctions that even the Bedouin nomads couldn't find. One story has it that an Air Force officer, operating under diplomatic passport while it was still possible, flew to Baghdad in late August 1990 with nothing more than a briefcase containing an SLGR. Once there, he was driven to the U.S. Embassy, and went to the courtyard to sit on a particular bench to wait for the GPS constellation to fly overhead (there were only six satellites up at that time) and take a single fix. Once this had been done, he got up, went back to the airport, and flew home with that one waypoint in the memory of his SLGR. From that one firm geographic fix came all of the targeting coordinates for the Tomahawk cruise missile and F-117 Stealth Fighter targets that were hit early in the war. Later on, so important was GPS to the conduct of the war that the famous "Hail Mary" sweep into Iraq could not have taken place without it.

Now that the system has been pretty much completed, the Army and the other services are rushing to put GPS receivers onto virtually everything that moves. Tanks, helicopters, fighters, guided missiles, and even trucks are all being equipped with the new navigational tools. In late 1993, the largest

GPS receiver procurement program running is the Portable, Lightweight GPS Receiver (PLGR—called a "plugger" by the troops) being procured from Rockwell International. These handheld receivers look like oversized portable calculators, with LCD displays and keyboards, and will be issued to infantry units, scout and special forces teams, and other units requiring GPS navigational capabilities. In addition, there are many other GPS-based programs being developed, all of which will be more accurate than the systems that they will replace. And maybe most important of all, the GPS system is being made available for all kinds of civil uses. Everything from civil surveying to blind airliner-landing systems are being tested. By the end of 1994, it is probable that GPS receivers tied to moving map displays (remember the one in James Bond's Aston-Martin?) will be available as optional equipment on civilian automobiles. GPS may be the most exciting technology that the military has introduced in years. What makes it even more interesting is that it is something we all can use. Those who conceived it so many years ago deserve our thanks for this new kind of public utility, which finally tells us where we are, and how to get where we want to go.

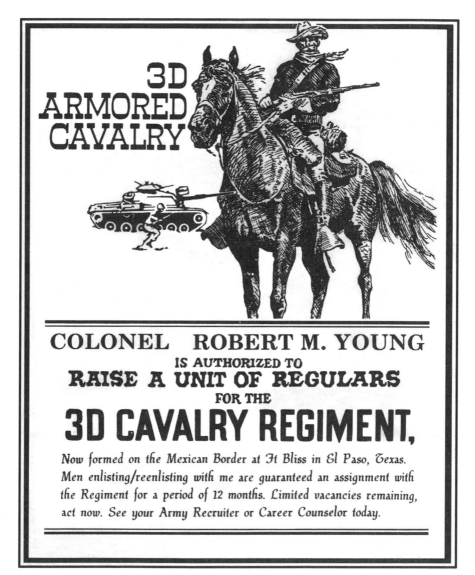

3rd ACR recruiting poster.

A Guided Tour of the 3rd Armored Cavalry Regiment

If you drive west from Fort Worth, Texas, along Interstate 20, past Abilene, Odessa, and the junction with Interstate 10, you come upon some of the most godforsaken country in America—West Texas. But then, as you drive up the Valley of the Rio Grande River, signs of life begin to reappear. And before long you are in the city of El Paso.

Just as El Paso is a crossroads between Mexico and the United States, it is also a crossroads between the past and the future in American history. Across the Rio Grande is the Mexican city of Ciudad Juarez, where Pancho Villa once crossed to raid American settlements. To the north are the old towns of Las Cruces and Alamogordo, New Mexico, which service one of America's spaceports at White Sands Missile Range. Nearby, Holloman Air Force Base is home for the F-117A Stealth Fighters of the 49th Fighter Wing. And nestled in between the past and future, on the north side of El Paso, is the old cavalry post of Fort Bliss.

Fort Bliss is the home of the 11th Air Defense Brigade, the soldiers who operated the Patriot air-defense system during the 1991 Persian Gulf War. But tucked over on the east side of the base is the home of America's last heavy armored cavalry regiment, the 3rd Armored Cavalry Regiment (3rd ACR). The 3rd ACR, assigned as the corps reconnaissance unit for the U.S. III Corps (based at Fort Hood, Texas), is one of the most powerful units for its size in the world. With lots of armored "teeth" and minimal administrative "tail," the 3rd ACR is a uniquely flexible and balanced combat unit—the smallest independent armored unit in the U.S. Army that might be deployed overseas to fight on its own.

Rapid deployment is critical to the American military. It took something like twenty-six shiploads to move General Barry McCaffrey's 24th Mechanized Infantry Division to Southwest Asia. With only eight of the high-speed (30+ knots/55 kph) SL-7 fast sealift ships in the Military Sealift Command's inventory, the 3rd ACR is the only heavy armor unit in the U.S. Army that can be lifted overseas on a single sortie by the SL-7 fleet. Such power and flexibility demand that the regiment's soldiers be among the best in the Army. When you meet them, that turns out to be the case. Let's take a look inside the 3rd ACR—first at its splendid history, and then as it stands in the summer of 1994.

The insignia of the 3rd Armored
Cavalry Regiment, the "Brave Rifles."
JACK RYAN ENTERPRISES, LTD., BY LAURA ALPHER

The 3rd ACR – A Short History

The 3rd ACR is the second oldest continuously established unit in the U.S. Army. Only the 2nd Armored Cavalry Regiment–Light (ACR-L) has a longer lineage. Created by an act of Congress on May 19th, 1846, the unit was officially assigned to securing a line of forts from Missouri to Oregon and maintaining order in the Western territories that were just starting to be settled. But the probable deeper reason for creating the regiment—to fight the Mexicans— was apparent from its equipment and size. Unlike traditional cavalry regiments equipped with carbines, sabers, and pistols, it was to be a regiment of mounted riflemen, a highly mobile fighting force with the punch of an infantry regiment. In addition, it was big. Organized just prior to the Mexican War, with billets for around 800 troopers, it increased the size of the active Army by fully 15%. The unit was officially organized as the Regiment of Mounted Riflemen on October 12th, 1846, at Jefferson Barracks, Missouri.

But before they could take up their stations on the Oregon Trail, the Mounted Riflemen were called to action in Mexico. The regiment fought in six separate campaigns during the Mexican War, including Vera Cruz, Contreras, Churbusco, and Chapultepec. During that action, on September 14th, 1847, they led the final assault that captured the Mexican National Palace, and thus earned the nickname they carry to this day, the "Brave Rifles." They returned to Missouri in 1848 to take up their duties on the frontier. Their first exercise was an epic six-month march to establish their presence in Oregon, arriving in November 1849. They returned to Missouri in 1851, and then moved on to Texas to suppress Indian uprisings and stop cross-border bandit raids from Mexico. At this time the regiment got its first look at its future home, Fort Bliss. Throughout the 1850s, the regiment was spread out in small bands from Texas to Arizona, and north to Colorado. It was tough, dangerous duty, dirtier and less romantic than portrayed by John Ford in a score of movies about the cavalry in the Old West.

The coming of the American Civil War in 1861 brought new challenges for the Brave Rifles. For starters, many of the officers and men joined the Confederacy. In spite of the leadership vacuum caused by the departure of the Southerners, the remaining officers concentrated the unit in New Mexico, where it was redesignated the 3rd U.S. Cavalry in August of 1861. The regiment remained in New Mexico until September 1862, when it returned to Jefferson Barracks, Missouri. During this time, it fought in both of the major battles of the war in the West. Moving on to Memphis, Tennessee, in December 1862, the 3rd did patrol and support duties until the following year, when it joined General Sherman's army during the drive on Atlanta. Following this, the Brave Rifles led the famous "March to the Sea" and the final drive into the Carolinas. Following the war, the regiment returned to the Western frontier, where it spent the next thirty years suppressing Indian uprisings and guarding the Mexican border. Of note was the Battle of Rosebud Creek (Montana) on June 17, 1876, the largest single engagement ever between the forces of the United States and Native Americans (1,400 troopers and friendly Indians against 4,000 to 6,000 Sioux and Northern Cheyenne), which was fought to a draw just a few weeks before the 7th Cavalry's debacle at Little Big Horn.

With the outbreak of war with Spain in 1898, the Brave Rifles deployed to Cuba, where they took part in the attack on San Juan Hill. Less than a year later, they again embarked to help suppress the Philippine Insurrection of 1899. The regiment spent the next couple of years fighting the rebels on Luzon, and eventually returned to Texas to fight border bandits in 1905. In 1917, the United States entered World War I. Sent to France, the 3rd did not see combat, but operated remount depots for U.S., British, and French horse-drawn units, returning to the United States in 1919. The Brave Rifles spent the next nineteen years based along the East Coast (mainly at Fort Meade, Maryland, and Fort Ethan Allen, Vermont), performing ceremonial honor guard duties in the Washington, D.C., area. At the outbreak of World War II, the regiment was still a horse cavalry unit. But a few months later it was re-equipped with scout cars and light tanks and redesignated the 3rd Armored Regiment, and after that the 3rd Cavalry Group. Deployed to Europe in 1944, the 3rd led General George Patton's drive across France, fought in the Battle of the Bulge, and joined the final drive into Germany.

Returning to Fort Meade following the war, it took on its present name, the 3rd Armored Cavalry Regiment, in 1948. It stayed in the U.S. until 1955, when it rotated to Europe, replacing the 2nd ACR until 1958, when it returned to Fort Meade. The Brave Rifles returned to Europe in 1961 during the Berlin Crisis, and stayed until 1968, when it again returned to the U.S., this time to Fort Lewis, Washington. Designated as a REFORGER unit, the 3rd ACR stayed at Fort Lewis until 1972, when it moved to its present home at Fort Bliss, Texas. The next eighteen years were spent in the routine but vital chores of a Cold War REFORGER unit.

The fall of the Berlin Wall and the death of Communism in Eastern Europe in the late 1980s might have meant the end of the line for the 3rd ACR. Like so many other proud Army formations, it was looking at having its colors cased and retired for good, until Saddam Hussein decided to invade Kuwait in August of 1990. On August 10th, 1990, the 3rd ACR was alerted to prepare for transport to Saudi Arabia to provide an armored reconnaissance element for XVIII Airborne Corps. By September of 1990, advance elements of the 3rd ACR began to arrive in Saudi Arabia to join Operation Desert Shield. Initially, their mission was to deter any further moves into Saudi Arabia by the Iraqis. But by early November, they knew that they were to be part of an army that would either force Iraq to vacate Kuwait, or evict it by force. As the first elements of the U.S. VII Corps began to arrive in December, the XVIII Airborne Corps started their move to the west as the flank covering force for the "Hail Mary" movement around the left side of the Iraqi line. On February 22, 1991, 3rd ACR led the rest of XVIII Airborne Corps through the berms into Iraq on their drive to the Euphrates River. During the four-day drive to the north, working closely with the 24th Mechanized Infantry Division, they covered some 183 miles/300 kilometers before they turned east towards Basra and the Rumaylah Oilfields. Their Desert Storm mission completed, they returned to Fort Bliss by April 5, 1991, to continue their cycle of training and modernization. As they head towards their 150th consecutive year of active service, they are likely to stay at Fort Bliss.

The 3rd ACR – Organization and Equipment

Since their return from Southwest Asia in 1991, the regiment has undergone a massive change in equipment and mission. Previously, the 3rd ACR was considered a reinforcing unit, with a minimal set of training equipment at home and their fighting equipment stockpiled at depots in Germany. Early in 1992, as part of the Army's reorganization for greater mobility, the 3rd ACR was redesignated as part of Force Package-1. Force Package-1, which includes units of III Corps and XVIII Airborne Corps, has been trained to the highest level of preparedness, and has the highest priority in equipment modernization. Suddenly, the Brave Rifles had to assimilate many new systems simultaneously. Everything from OH-58D Kiowa Warrior scout/attack helicopters to the new SINCGARS family of radios appeared on 3rd ACR's Table of Organization and Equipment (TO&E)—all in about twenty-four months. This is a challenge that will test every member of the regiment to the limit. But when the current modernization cycle is completed around 1996, the 3rd ACR will be the most powerful ground unit for its size in the world. Let's take a look at the people and their tools.

Before we get started, though, a quick note about unit designations. The cavalry has always been a community unto itself, and this tradition of doing things its own way continues even today. Thus, while the basic Army building block, the platoon, remains the same for the cavalry and armor communities, the next level up the chain is different. What an armor officer would call a company (usually between three and five platoons), the cavalry officer calls a troop. Similarly, what an armor officer calls a battalion (four to six companies), a cavalry officer would call a squadron. Thus, the terms "company" and "troop" are equivalent, as are "battalion" and "squadron." With that detail clarified, let's look at the 3rd ACR.

Headquarters Troop

The Regimental Headquarters and Headquarters Troop (RHHT) includes the regimental commander, a colonel, and his immediate staff and some command and support vehicles. These include:

- A pair of M3A2 Bradley cavalry vehicles configured as "command tracks." Usually equipped with two or more radios, these are otherwise standard Bradleys.
- Eleven to fifteen M577 Command APCs, organized into groups of four or five. These will be replaced by the new XM4 command vehicle when it comes on-line later in the 1990s.
- A number of HMMWVs and trucks to provide general transportation and support.

The command sergeant major (the senior NCO of the regiment) and the regimental chaplain, as well as some clerks and support troops, complete this compact headquarters.

In late 1993, the regimental commander was Colonel Robert Young. A 1969 graduate of Texas Christian University, he is a career cavalry officer, with numerous tours in the 2nd ACR while they were still based in Germany. In addition, he has an extensive background in Middle East relations, holding a masters degree in the subject. Thus, when he has not been leading cavalry troops, he has been supervising operations like the withdrawal of Israeli forces from Lebanon (as a UN observer), and acting as the U.S. representative to the UN relief effort for the Kurds in northern Iraq (Operation Provide Comfort). He must have done a good job there, for when he left Iraq at the end of 1992, Saddam Hussein had a bounty of $100,000 on his head. It is my understanding that the bounty still stood in late 1993. Colonel Young took command of the 3rd ACR in May of 1993, and will remain in this post until sometime in mid-1995. Working directly with Colonel Young is the regimental command sergeant major (RCSM), Dennis E. Webster. CSM Webster joined the Army in 1972 and is a career armor and cavalry NCO. He has been with the 3rd ACR since 1990, as command sergeant major (CSM) for the regiment's Support Squadron, 3rd Squadron, and now as the 12th RCSM of the regiment. The RCSM is the senior enlisted soldier in the regiment, and provides a direct link from the enlisted personnel to the regimental commander. The regimental executive officer (RXO) is Lieutenant Colonel Luke Barnett, and he supervises the headquarters staff. There is a small regimental staff consisting most notably of the regimental operations officer (S-3) and staff, who are responsible for planning and directing the way the regiment operates and fights.

1st "Tiger" Cavalry Squadron

Headquarters & Headquarters Troop (HHT)

"A" Cavalry Troop   "B" Cavalry Troop   "C" Cavalry Troop

"D" Tank Company   "D" Artillery Battery

The basic organization and equipment of the 3rd Armored Cavalry Regiment's 1st Cavalry Squadron ("Tiger" Squadron) commanded by Lieutenant Colonel Toby Martinez, USA. The 2nd and 3rd Cavalry Squadrons have the same basic layout. The M109A6 Paladin self-propelled howitzers are planned to replace the older M109A2s currently fielded by the Brave Rifles.

Jack Ryan Enterprises, Ltd., by Laura Alpher

1st, 2nd, and 3rd Cavalry Squadrons

The cutting edge of the regiment is the three armored cavalry squadrons, numbered 1st, 2nd, and 3rd. On paper, an armored cavalry squadron resembles a reinforced tank battalion, with a strength of 53 officers, 339 NCOs, and 499 enlisted troopers, for a total of 891 billets. Each squadron is commanded by an armor lieutenant colonel, with a command sergeant major and a small staff to help run things. In late 1993 the leaders of the three squadrons were:

- 1st Squadron—Lieutenant Colonel Toby W. Martinez (Murray State University, 1975) is the commander; and the senior NCO is CSM Roy Thomas.
- 2nd Squadron—Lieutenant Colonel Norman Greczyn (West Point, 1972) commands, with CSM Alton B. Eckert as the command sergeant major.

Cavalry Troop

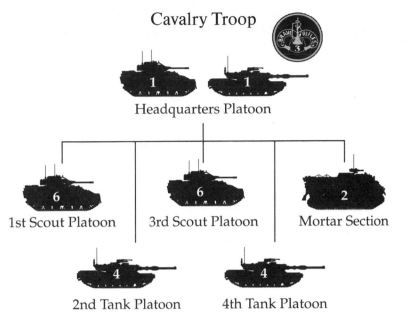

Headquarters Platoon

1st Scout Platoon   3rd Scout Platoon   Mortar Section

2nd Tank Platoon   4th Tank Platoon

The equipment and organization of a cavalry troop. The 3rd Armored Cavalry Regiment has a total of nine such troops in its table of organization and equipment (TO&E).

JACK RYAN ENTERPRISES, LTD., BY LAURA ALPHER

- 3rd Squadron—Lieutenant Colonel Karl J. Gunzelman (West Point, 1975) commands, with CSM Conrad C. Bilodeau as the ranking NCO.

Each squadron is a self-contained combined-arms team consisting of:

- Headquarters and Headquarters Troop—Two M3A2 Bradley command tracks, six M577 command vehicles, some HMMWVs, and various recovery vehicles, trucks, trailers, and fuelers for all types of combat service support.
- Three Armored Cavalry Troops—These make up the cutting edge of the squadron. Each troop consists of an HQ section (one M1A1 Abrams tank, one M3A2 Bradley, and one M577 command post), two scout platoons (six M3A2 Bradley cavalry fighting vehicles per), two tank platoons (four M1A1 Abrams tanks per), a mortar section (two M106 4.2"-mortar carriers), and twelve supporting tracked and wheeled vehicles.
- Tank Company—Each squadron has a pure tank company (never referred to as a "troop") to provide an armored reserve for the squadron commander. It consists of an HQ section (two M1A1 tanks) and three tank platoons (four M1A1 tanks each).
- Howitzer Battery—To give the squadron its own organic artillery support, it is assigned a battery of eight M109A2 self-propelled howitzers (SPHs), with eight FAASVs in support.

The commander of an armored cavalry squadron leads forty-one M1 Abrams tanks, forty-one M3 cavalry vehicles, eight M109 SPHs (with eight FAASVs), and six M106 4.2"-mortar carriers. This is a lot of firepower at one leader's fingertips! Remember that just one armored cavalry squadron wrecked an Iraqi Republican Guards brigade (three to four times the size of the American force) at the Battle of 73 Easting during Desert Storm. And in the next few years, the squadrons will enhance their combat power with the new M1A2 Abrams, the M109A6 Paladin SPH, the IVIS version of the M3A2 cavalry vehicle, and the new 120mm-mortar carrier.

Within the regiment, each troop (which corresponds to a company-sized unit) is commanded by a captain and designated by a letter of the alphabet. The First Squadron includes A, B, and C Troops, and D Company; the Second Squadron has E, F, and G Troops, and H Company, etc. By tradition, each troop adopts a name based on its identifying letter. For example, E Troop is "Eagle," and I Troop is "Ironhawk." Overall, the current cavalry squadron structure provides massive gun and missile firepower and built-in supporting arms. If it has a weakness, though, it is the lack of dismountable infantry (each M3A2 has only two dismount scouts per vehicle; these are *not* infantry soldiers), which limits its ability to hold and clear terrain. On the other hand, the cavalry can cover ground and dish out punishment like no other unit of equivalent size in the world. We'll look at some typical armored cavalry operations in the next chapter.

4th Air Cavalry Squadron

The 4th Air Cavalry Squadron is commanded by an aviation lieutenant colonel and has 36 officers, 110 warrant officers and NCOs, and 355 enlisted troopers, for a total strength of 501. In late 1993, the commander was Lieutenant Colonel Gratton Sealock (Eastern Washington State College, 1974), with 1st Sergeant Timothy D. Paul as the CSM. The unit includes:

- Headquarters and Headquarters Troop—The HHT includes three UH-60L Blackhawks, three EH-60 "Quick Fix" electronic-warfare helicopters, and one OH-58C Kiowa scout helicopter. The support element is similar to its ground squadron counterpart.
- Three Air Cavalry Troops—Each of these troops consists of six scout helicopters (currently OH-58Cs) and four attack helicopters (currently AH-1Fs).
- Two Attack Helicopter Troops—Each of these troops consists of four scout helicopters (currently OH-58Cs) and seven attack helicopters (currently AH-1Fs).
- One Transport Helicopter Troop—This troop consists of fifteen of the newest-model UH-60L Blackhawk helicopters.
- Maintenance Troop—This is a ground-based troop that supplies maintenance and support to the rest of the 4th Squadron.

In total, the 4th Squadron operates seventy-four helicopters (twenty-six attack, twenty-seven scout, eighteen transport, and three electronic warfare helicopters). As it is currently equipped, the air squadron has both some of the newest and oldest equipment in the Army. On the positive side, the UH-60L Blackhawks and EH-60 "Quick Fix" birds are state-of-the-art, and capable of any mission that might be assigned to them. A particular asset is that the Blackhawks are equipped with the new ESSS stores system, for carrying extra fuel, and even Hellfire missiles (which require another unit to laser-designate targets for them.)

The 4th Squadron's big problem is that the current force of OH-58C Kiowa scout helicopters and AH-1F Cobra attack helicopters is obsolete. These are incapable of dedicated night operations, laser designation, automatic handoff of targets, or long-range stand-off missile fire (the AH-1 is only equipped with TOW missiles). This is a serious shortcoming in a unit so likely to be deployed in response to a fast-breaking crisis. So weak is the firepower in the old squadron structure, that when 2nd and 3rd ACR deployed to the Persian Gulf during Desert Shield, both had a battalion of AH-64A Apache and OH-58D Kiowa Warriors attached to make up for the limitations of the unit's equipment.

Like the tankers and gunners in 1st, 2nd, and 3rd squadrons, with their dreams of M1A2s and M109A6s, the aviators of the 4th Squadron look forward to Apaches and Kiowa Warriors to replace their old birds. When this will happen is uncertain, mainly because the production run for new AH-64s has been terminated, and the OH-58D is being produced year-to-year at the discretion of Congress—and the strength of the Texas congressional delegation. Because the Army prefers to put money into the Comanche program, it has not

4th Air Cavalry Squadron

Headquarters & Headquarters Troop

"M" Aero Scout Troop   "O" Aero Scout Troop   "Q" Attack Helicopter Troop

"R" Transport Helicopter Troop

"N" Aero Scout Troop   "P" Attack Helicopter Troop

The organization of the 4th (Air Cavalry) Squadron as it will appear when it finishes its modernization program. The 3rd Armored Cavalry Regiment attack and aero scout troops are currently equipped with relatively obsolete AH-1F Cobra attack helicopters and OH-58A/C Kiowa scout helicopters.

Jack Ryan Enterprises, Ltd., by Laura Alpher

3rd Armored Cavalry Regiment

M1A2 · 123

UH-60L · 21

M3A2 · 127

M109A6 · 24

OH-58D · 26

Mortar Carrier · 18

AH-64 · 26

The types and numbers of different weapons systems that the 3rd Armored Cavalry Regiment should be equipped with when it completes its current modernization cycle. *Jack Ryan Enterprises, Ltd., by Laura Alpher*

asked for any new OH-58Ds. The 4th Squadron will be modernized eventually, though when and how remain a mystery.

Support Squadron

The Support Squadron is commanded by a lieutenant colonel and has a total strength of 802 personnel. The current commander is Lieutenant Colonel Thomas M. Hill (University of South Dakota, 1972), and his command sergeant major is CSM Halford M. Dudley. Lieutenant Colonel Hill has a delightful sense of humor. This is a great asset, for he faces a daunting task: supplying everything needed to keep the 3rd ACR moving, fighting, and living. This means everything from food to floppy disks, dental care to bulldozer repair. If it is in the regiment, then Lieutenant Colonel Hill and his soldiers will keep it going. Though not much larger than a cavalry squadron in numbers, the Support Squadron has the most diverse assets in the regiment. It includes:

- Headquarters and Headquarters Troop—This is the nerve center of the squadron, with a heavy base of data-processing personnel, a few trucks, and HMMWVs.
- Medical Troop—Equipped with sixteen ambulances (eight HMMWV-based, eight M113 APC-based), the medical troop is staffed to set up aid stations to process casualties to divisional/corps-level MASH-type hospitals.
- Maintenance Troop—The automotive maintenance troop is tasked with providing direct-support recovery and maintenance for the

A 3rd ACR M88 tank recovery vehicle in the Fort Bliss tank park. These vehicles are used to tow or recover damaged or disabled armored vehicles. *JOHN D. GRESHAM*

regiment. Equipped with a variety of trucks (twenty-two five-ton tractors and twelve five-ton trucks) and recovery vehicles (five M88, four 5-ton wreckers, and three M113 maintenance trucks), it can service any vehicle in the regiment. It repairs equipment that mechanics at the troop and squadron level do not repair.

• Supply and Transportation Troop—The supply troop is just what it sounds like, a combination of warehouse and trucking company. Equipped with eighty-seven heavy trucks (thirty-three 5-ton tractors, twenty-six 5-ton trucks, twenty-two 5,000-gallon/18,868-liter HEMTT tankers, and six HETs), it moves something over 559 short tons/508 metric tons of cargo and water a day for delivery to the front-line units of the regiment.

It is difficult to explain just how much it takes to run an ACR in the field for a day. The numbers just begin to numb you after a while. For example, the support squadron is equipped to purify and distribute 139,750 gallons/527,358.5 liters of water a day. Based upon the squadron having to service 5,000 personnel at a time, that's about 9.6 gallons/36.2 liters a day for every soldier in the regiment (for washing, drinking, food preparation, personal hygiene, etc.). Then there is food. The standard five-day field-ration issue (part of what is called a Unit Basic Load or UBL) for the regiment is 75,000 MREs (about 15,000 per day), in 6,250 cardboard cases, strapped onto ninety-eight pallets, weighing just over 75 short tons/68 metric tons. That is just food for five days in the field. And then there is fuel (200,000 gallons/754,716 liters of diesel fuel, 50,000 gallons/188,679 liters of JP-5 jet fuel), clothing (275 items), spare parts (2,793 different types for the ground equipment, 3,361 for the aviation equipment), and everything else that keeps this movable equivalent of a small town working.

It is a huge job. That it is done is a masterpiece of organization, data processing, communications technology, and a lot of hard work by soldiers who rarely get medals for their efforts. Their efficiency would amaze most taxpayers, whose image of the Quartermaster (supply) Corps is based on the stereotype of the wheeling-dealing supply sergeant. There may still be a few of those guys around, but the Army is moving into a new era of "just in time" mainte-nance planning and "as you need it" supply deliveries. Previously, for example, each tank platoon kept a pair of spare generators for the M1 tank in its platoon stocks. Now they do not have any on hand at all. When they need one, the troop supply clerk types a request on a laptop computer, and in a few hours, the part is delivered from a centralized stock at the regimental level. Meanwhile, the regimental supply system will have automatically ordered another genera-tor over the world-wide Army supply computer network, so that by the time the generator is being installed on the tank back at the platoon, the replacement generator for the regimental stores is probably being packaged for shipment by Federal Express (yes, they even go to combat zones!) for immediate delivery. In this way, the overall Army-wide stock for an item can be reduced, and the tax-payer saves money. It also makes the ACR a leaner (not having to carry so much of a "tail"), meaner (more space for fuel and ammunition) unit.

During Desert Shield/Desert Storm Lieutenant General "Gus" Pagonis managed the logistical effort for CENTCOM in the Persian Gulf. When he later wrote a book on the subject, he titled it *Moving Mountains*. Never has a book carried a more apt title, for what the people of the Army Quartermaster Corps and the soldiers of the 3rd ACR supply squadron do is just that, move mountains.

43rd Combat Engineer Company

The 43rd Combat Engineer Company is commanded by an engineer cap-tain, and has a strength of 220 soldiers. It has its own maintenance platoon, an assault and barrier (A&B) platoon, and three identical combat engineer pla-toons. Its TO&E includes:

- Twelve M113 armored personnel carriers
- Six M9 armored combat earthmovers
- Three armored vehicle-launched bridges (an M60 tank chassis with a folding scissors-bridge for crossing small streams and big ditches)
- Three combat engineer vehicles (an M-60 tank chassis with a dozer blade and a short-barreled 165mm "demolition gun")
- Six 5-ton dump trucks
- One scoop loader
- Assorted other specialized excavating and entrenching vehicles

The mission of combat engineers is to clear away obstacles that impede friendly movement and to create obstacles to impede enemy move-ment. Engineers are trained for demolition, construction of field fortifica-tions, and repair of roads and bridges. They are particularly vital for combat in urban areas.

66th Military Intelligence Company

The Military Intelligence (MI) company is commanded by a military intelligence major and has a strength of 152 personnel. It operates a variety of electronic jamming and surveillance systems.

89th Chemical Company

The chemical company is commanded by a Chemical Corps captain, and has a strength of 78 troopers. Although all civilized nations have signed treaties renouncing the use of chemical weapons, the Army's chemical troops have an important protective mission (not all our potential enemies are civilized). The Chemical Corps also has a traditional mission of laying smoke screens. The company's TO&E includes:

- Six Fox NBC reconnaissance vehicles
- Seven M1059 smoke generators (based on M113 chassis)
- One M12A1 decontamination apparatus
- Assorted other specialized detection and decontamination equipment

In operations under a chemical threat, the 89th assigns its Fox vehicles, usually two per cavalry squadron, so that there is a highly mobile chemical-detection-and-survey capability up front at all times. Should an area of contamination be encountered, the Fox can quickly find a bypass route, keeping the squadrons moving and uncontaminated. This prevents other units from getting "slimed" by running through a contaminated area. Normally a corps decontamination platoon will be added to the 89th, so it can handle the job of equipment and personnel decontamination.

Air Defense Section

The 3rd ACR has an air defense section attached to the Regimental Headquarters and Headquarters Troop. The section consists of six Avenger air-defense vehicles. These are Hummers mounting compact turrets that combine a .50-caliber machine gun, eight Stinger missiles in a pair of canister launchers, and a digital fire-control system with a laser rangefinder, a thermal sight, and a FAADS early warning terminal. Avenger will be replaced in a few years when the Bradley Stinger vehicle is fielded. The section may also be able to deploy shoulder-fired Stinger teams with their own HMMWVs.

U.S. III Corps—Attachments and Contributions

As we mentioned earlier, the 3rd ACR is one of a number of units attached to the U.S. III Corps at Fort Hood, Texas. III Corps includes:

- 1st Cavalry Division (actually an armored division)
- 2nd Armored Division
- 1st Mechanized Infantry Division
- 4th Mechanized Infantry Division
- 3rd ACR
- Attached field artillery, air defense, engineer, military police, and support units

It is the single most powerful ground force in the world today. Commanded in late 1993 by Lieutenant General Paul Funk (who commanded the Third Armored Division during Desert Storm), III Corps would decide how to employ 3rd ACR for a particular mission, and how it would be augmented with appropriate support units.

One of the things that makes an ACR such an attractive unit to send overseas is the ability to custom tailor it for diverse mission requirements. For example, consider the situation faced by Colonel (now Brigadier General) Don Holder, who commanded the 2nd ACR during Desert Storm. The 2nd ACR was organized exactly as the 3rd is today. To support his mission, a movement-to-contact with several Iraqi Republican Guard armored divisions, the VII·Corps commander, then-Lieutenant General Fred Franks, added a number of different units to beef up the combat power of the regiment. These attachments included:

- 210th Field Artillery Brigade (M109s and MLRS)
- 2nd Battalion/1st Aviation Regiment (OH-58Ds and AH-64As)
- 82nd Combat Engineering Battalion
- 214th Military Police Company
- 177th Personnel Services Company

With all these additions, 2nd ACR resembled a small armored division, and not the lean cavalry regiment that had come ashore at Al Jubail. But these were things that the corps commander felt they might need, and the open structure of the cavalry regiment allowed this to be done easily.

So, what might Lieutenant General Funk provide to Colonel Young if he had to go off to combat in a distant land, where the situation was uncertain, and the goals still being kicked around the White House Situation Room? A helicopter squadron such as the 4th Squadron of the 6th Cavalry Brigade at Fort Hood, Texas, would be a good start. Equipped with Apaches and Kiowa Warriors, it would give Colonel Young the night-surveillance capability and long-range missile punch he currently lacks. Another candidate would be additional artillery, perhaps a battalion of MLRS launchers from the III Corps Artillery at Fort Sill, Oklahoma. Add the 1st Military Police Company (from Fort Riley, Kansas), to help with rear-area security, POW processing, and traffic management. Add all of this up, and a potential foe might decide not to even try to make trouble. Battles you prevent can be victories even sweeter than the battles you win.

Honing the Razor's Edge

In many cultures, during much of history, soldiers were seen as the dregs of society, and were recruited from the gutters. An ancient Chinese saying, "Good iron is not used for nails, and good men are not used for soldiers," sums up this traditional attitude of emperors, kings, and perhaps even a few presidents. But not so in the American military tradition, which is based upon the volunteer citizen soldier and unquestioned civilian control over the Armed Forces, exercised by the people's elected representatives. In our tradition soldiers most decidedly do not come from the dregs of society.

Though for long periods of American history, the professional officer corps has been a small, inbred, inward-looking aristocracy, in times of war, Americans have usually managed to field a people's Army, led by popular generals such as Omar Bradley and Ulysses S. Grant. The simple truth is that America's greatest field forces—from Washington's Army at Yorktown to George Patton's Third Army in Europe—have always been made up of trained citizen soldiers, motivated by American ideals. It is the American style of war.

So what does America's Army look like today? Well, very much like America itself. It is increasingly an African-American, Hispanic, Asian-American, and Southern Army, because those are the growing demographic sectors of our population. But the Army also recently commissioned its first Islamic chaplain, because there is a growing segment of Muslims in America (and at this writing, the Army is still looking for its first Buddhist chaplain). It is also an Army where women are increasingly visible in non-traditional and leadership positions, because we are a society where women have struggled for over a century to win greater opportunities. Women are now being admitted into combat arms, such as Aviation and Artillery, without *any* gender-based restrictions.

In many ways, our Army struggles with the same problems of stress, family breakdown and separation, and alcohol abuse as every other community. But remarkably, it is an Army that is virtually drug-free, thanks to a rigorous program of random testing. Perhaps equally remarkable (but maybe not, when you think about it), in a period when traditional religions are losing believers, we have perhaps the most religious Army since Robert E. Lee led the Army of Northern Virginia in prayer before battle. In a period of high unemployment and declining numbers of entry level jobs, it is also a shrinking Army. This is a real challenge for the Army's recruiters, its advertising agencies, and its public-relations staff. By 1996, the Army will have reduced-in-force (prematurely retired, involuntarily separated, laid off, or whatever you want to call it) a number of soldiers equal to the number it sent to fight in the Persian Gulf War of 1991. How do you convince the best and brightest young people in our society that there are great opportunities in a downsizing organization with old-fashioned values and the risk of sudden, violent death?

Why *do* people join the Army? I've asked this question as I traveled around the country researching this book. Some of the answers include:

- Educational, travel, and training opportunities
- Recruiting/re-enlistment bonus money
- Family or community traditions
- A sense of adventure or patriotism
- A sense of belonging

For some, the Army is a path up from the gangs and violence of the inner city or the despair of poverty. For many it is an opportunity to make their own way in the world. All of these are reasons for young men or women to consider the Army as a place to start their adult lives or make a career. That has been the attraction of the Army for many men and women of all races, religions, and backgrounds. It is an organization that looks like the country that it protects, serves, and frequently represents to the rest of the world. It therefore is no surprise that when Americans are asked who they respect and trust most, several well-known ex-Army officers are at the top of the list.

The Enlisted Troops

To maintain its projected active-duty strength of around 500,000 personnel, the Army still needs over 100,000 new enlisted recruits per year.

Suppose that you've just graduated from high school (you have to be at least seventeen years old to enlist), and you drop in to see your local Army recruiter. (If you have a drug problem, or a police record for anything more serious than minor infractions, forget it.) The recruiter will ask what kind of training and career specialty you want to pursue in the Army, and describe all of the options available. There are a lot of them, and this may take some time if you have not thought out exactly what you want. If you check out okay, you will be asked to sign some papers, similar to a contract, in which you agree to enlist for a certain term of years (this varies with the chosen specialty—check out the current rules). Then you will be given a medical exam and processed for induction. There is a small ceremony in which you take an oath to "preserve, protect, and defend the Constitution of the United States against all enemies, foreign and domestic."

After induction, new recruits report for Basic Training, which lasts twelve weeks. Basic Training centers include Fort Knox in Kentucky, Fort Jackson in South Carolina, Fort Leonard Wood in Missouri, Fort Lee in Virginia, Fort Benning in Georgia, and a number of other posts. From Hollywood movies, most people get the impression that Basic Training (better known as "Boot Camp") is a cross between a prison chain gang and a Nazi concentration camp. But in practice, the Army has learned that harassment and brutality simply don't work, especially with intelligent and motivated recruits. All the same, Basic Training is designed to be physically demanding and psychologically stressful, yet it is also designed to build small-unit cohesion, fitness, and self-esteem, along with some of the skills of soldiering.

The Army's approach to training is centered around three principles: Task, Conditions, and Standards. Soldiers learn their jobs as a series of tasks, with conditions to be met, to a certain standard of performance. Thus, a typical training Task might be stated as:

> "After completing this lesson, soldier will be able to put on a Nuclear-Biological-Chemical (NBC) protective mask."

The Conditions of the task might be:

> "He will be given an M49 mask in its carrying case, in the presence of simulated chemical agent (yellow smoke or in some cases, CS tear gas)."

The Standard might be:

> "to don the mask within 5 seconds, and without any detectable gaps or leaks in fit."

The Army devotes tremendous effort to designing, developing, testing, and evaluating its training methods and materials. Soldiers want realistic hands-on training, and when this is not possible for reasons of cost or safety, it is often possible to provide simulators, mock-ups, and similar training devices. Standards are intentionally set high because for all the excitement and adventure of military service, it is a *dangerous* profession, where people can get hurt using equipment improperly. Army instructors are taught that if the student fails to learn, it is because the instructor *failed* to teach properly.

After graduating from Basic Training (if you fail the first time, you can cycle through it one more time), soldiers are assigned to Advanced Individual Training (AIT) in a particular branch or specialty. The branches of the Army fall into three main groups:

- **Combat**—Infantry, Armor, Artillery.
- **Combat Support**—Aviation, Air Defense, Artillery, Engineers, Military Police, Military Intelligence, Signal, etc.
- **Combat Service Support**—Quartermaster, Transportation, Finance, Data Processing, etc.

At the completion of AIT, which typically takes from six to twenty-four weeks, the soldier receives his Military Occupational Specialty (MOS) Code. If you want to be an Armored Cavalry Scout, the MOS is 19D (pronounced Nineteen-Delta), and you would attend the Armor School at Fort Knox, Kentucky. If you want to be an Apache Helicopter Mechanic, the MOS is 67R (Sixty-Seven-Romeo in the vernacular of the trade), and you would attend the Aviation Maintenance school at Fort Rucker, Alabama. When soldiers graduate from AIT, they are given their first assignment and sent off to their first unit.

Enlisted soldiers advance through a series of nine grades of rank, starting with E-1 (private) through E-9 (sergeant major). Senior enlisted personnel are highly respected in the Army; it is not unusual for a wise officer to ask the senior NCO's advice on tactical situations, or how to deal with a problem soldier. Such personnel are frequently college-educated, and you often find sergeants major (yes, that's the proper plural form) with graduate degrees.

From the time of his first assignment, the life of an enlisted person can be summed up in just a few words: do your job, go to school, and prepare yourself for the next rank up the chain. Because of the Army's force drawdown, the life is becoming more competitive. In a time of declining force levels, only those showing skill and promise are going to be retained.

On the other hand, there are jobs for Army enlisted troops that are simply not available in any other branch of the military. For example, if you want to fly as a pilot in the U.S. Air Force, you have to be an officer. Not so in the Army. There is a tradition of "flying sergeants" going back decades; and when the Army decided to maintain an aviation arm after the Army-Air Force split in 1947, they decided to continue the tradition as a series of ranks for warrant officers (WOs). Warrant officers are former sergeants, with their own separate career path for jobs of high responsibility—such as the care and operation of a ten-million-dollar helicopter! The Army trusts enlisted soldiers with jobs just as important, responsibilities just as great, and objectives just as vital as those given to officers. In fact, the only thing an officer can do that an enlisted soldier can't is command.

The Officer Corps

They range from cabinet-level officials in the government to second lieutenants commanding platoons, and yet they all have one thing in common: the trust of a nation to use their best judgment to fulfill their duty. That is a big responsibility, as the last few years have shown. The Army expects a lot from its officers.

Just before the 1991 Persian Gulf War, it was top Army officers in Washington D.C., particularly General Colin Powell, the Chairman of the Joint Chiefs of Staff, and General Carl Vuono, the Chief of Staff of the Army, who briefed the Bush administration on how a war against Iraq to liberate Kuwait could be fought and won with acceptable risks and casualties.

Meanwhile, on the other side of the world, some of the Army's most junior officers were preparing to lead their soldiers into combat. It was, in fact, three young Army officers, Dan Miller, H. R. McMaster, and Joe Sartiano, each leading a troop from the 2nd Squadron of the 2nd ACR, who made first contact with the security elements of Iraq's Republican Guard, fighting the decisive ground action of the Gulf War. Three captains, not one of them more than twenty-eight years of age, made on-the-spot judgments that determined where the rest of General Franks' VII Corps would fight. What does General Franks think about the quality of their initiative today? The three young officers "did exactly what I would have done," he says.

What a difference from a quarter century ago, when Army junior officers were regarded as the least professional officers in the U.S. military—and some were "fragged" (shot in the back by their own troops)!

That difference is due to the intellectual and professional growth of the commissioned officer corps of the Army. Note that word *commissioned*. A soldier is *enlisted* in the Army for a certain term of years, and may be offered the opportunity to re-enlist (much like a contract). But an officer is *commissioned*. The commission of an Army officer asks many things of the person accepting it. These might best be summed up by the motto of West Point: "Duty, honor, country." Depending on performance and the needs of the service, an officer may hold that commission until the age of retirement. But let's look at the start of the journey.

An Army officer's career usually begins after high school, when a young person decides to join up. Then comes college, an experience which is of great interest to the Army. And the interest is not passive either. For it can help a teenager afford a college education in a number of ways. These include:

- Appointment to the U.S. Military Academy at West Point, New York.
- Reserve Officer Training Corps (ROTC) scholarship programs at major colleges and universities.

From these programs, as well as from Officer Candidate Schools (for young enlisted soldiers who already have college degrees), come second lieutenants (also known as "butter bars" from the golden color of their insignia of rank), the raw material of the Army officer corps.

From the start of his or her career, the officer has one goal—move up or move out. Officers can only be passed over for promotion two times before they are forced out of the Army. And currently, lieutenants trying for promotion to captain only get *one* try. Like their enlisted counterparts, they face the problem of a declining force size in the Army.

Since being qualified to do a lot is the best way to stay in, the post-Vietnam generation of Army officers is the best-educated officer corps that America has ever put into the field. It is expected that in the course of a twenty-year-plus career, an officer will obtain a graduate degree and constantly seek out professional educational opportunities. Some of the more interesting of these are:

- The Army War College at Fort McNair in Washington, D.C.
- The Command and General Staff College at Fort Leavenworth, Kansas.
- The School of Advanced Military Studies, the graduates of which are known as "Jedi Knights."
- Specialized training schools, duty as exchange officers, and advanced degrees from civilian universities.

The result of all this training is that as a group, the officers in the U.S. Army today are the most professional in the country's history. Many are also, of course, veterans of combat in the Persian Gulf, Panama, Somalia, and other places. Combat experience tests the war-fighting skills and effectiveness of a military force. Other things help, but the victorious combat experience of those young captains and majors will help maintain the Army's war-fighting skills for another generation. And that will be good for everyone they command.

Command is the ultimate goal of a professional officer. Luckily, the Army's organization allows a large percentage of the officer corps to gain early command experience as platoon and company/troop leaders. It also gives the Army an opportunity to evaluate them: Based upon their performance as commanders of small units, it selects those best qualified for command of larger units. This does not necessarily guarantee that the best officers always rise to the top, but it does tend to push the ones with talent and potential into positions of responsibility, where their management skills, initiative, and leadership under stress can be fairly evaluated.

The Road to the National Training Center

The goal of any Army combat unit is to be ready to deploy and, if necessary, be ready for combat. How does a commander like Colonel Young of the 3rd ACR take the newly schooled soldiers sent to him by the Recruiting Command, mix them with the equipment and soldiers he already has, and make them combat-ready? The Army has a habit of rotating and promoting people in such a way that maintaining readiness is a constant challenge. Colonel Young's responsibility as the regimental commander is to maintain the fine instrument left to him by the previous commander, Colonel (now Brigadier General) Robert R. Ivany. Luckily, the Army has a whole series of training opportunities and exercises designed to help their unit commander do just that. This training cycle is set up so that at the end of it—the "final exam"—the whole regiment gets to find out just how well it has done. This final exam is called the National Training Center (NTC).

The National Training Center Concept

NTC was created because of the generally poor performance of U.S. Army units in Vietnam. It is designed to be a force-on-force training environment where units up to regiment and brigade size can maneuver and fight in a simulated war zone for a period of weeks. Located at Fort Irwin, California, in the Mojave Desert (near Barstow, California, south of Death Valley), NTC gives Army units the chance to do their fighting in a controlled environment. The concept for NTC came from a study of early combat experience that indicated that soldiers and units in action for the first time suffered the worst casualties. For example, the Navy found that if a pilot survived his first ten missions over North Vietnam's formidable air defenses, he was much more likely to survive the next ninety. The studies indicated that the stresses of combat and the confusion of battle tended to make young pilots almost inept for a time, until they learned to create the "mental filters" that allow a combat veteran to distinguish what is critical to survival from what can safely be ignored. Pilots call it "situational awareness," and it is the characteristic that separates an ace from a corpse. The Navy's answer to this problem was the creation of the famous Top Gun fighter weapons school, and later the Strike University at NAS Fallon in Nevada. So successful was the Top Gun program that the Air Force opened a similar schoolhouse for warriors at Nellis AFB, Nevada, under the name Red Flag.

The Army also realized the benefits of such a program; and at the same time it wanted to create a training center to teach the art of the new maneuver-style warfare that was emerging as its standard doctrine. The wide Mojave Desert in California was an obvious location, and Fort Irwin, outside Barstow, California, became available in the early 1980s. Fort Irwin was a vast, decrepit old post, almost unused since General Patton trained armored units there in the 1940s. It took years of work and lots of taxpayer money to build up the facilities; and even now construction (particularly of base housing) continues. What makes Fort Irwin such a perfect place for practicing maneuver warfare is summed up in one word: space. The NTC complex at Fort Irwin covers about 1,000 square miles/3,050 square kilometers (about the size of Rhode Island), all completely open and government-owned. Another advantage is that no one cares much about the place. It is as barren a desert as America has to offer, and operations there are unlikely to disrupt civilian activities. Nor are there natural features that outsiders care about—except for the world's largest desert tortoise hatchery and a rare species of brine shrimp inhabiting the seasonal dry lakes of the region (and the Army *is* concerned about protecting these). Otherwise, it is a really big and dusty playing field. Though temperatures in the summer are extreme, rising to over 110°F/44°C regularly, still, all that room makes it the perfect sandbox to practice the art of war. Resident at the NTC is a simulated Soviet-style motor-rifle regiment staffed by Army soldiers called the Opposing Force (OPFOR). And the entire NTC facility is instrumented, to allow recording and playback of entire battles. A unit and its commanders can therefore learn exactly what they did wrong (and right) in battle against the OPFOR.

Units deploy to Fort Irwin for three-to-five-week "rotations," which are designed to test basic combat skills of gunnery and maneuver, plus supporting skills, such as logistics, combat medicine, and maintenance. All of this is designed to be a pure twenty-four-hours-a-day learning experience that encourages participants to do new things and be innovative, as well as teaching officers and troops to maneuver in a realistic environment. All of this is obviously very expensive; yet the experience gained teaches lessons that cannot be simulated on a computer, or played out at a unit's home base. More important, the experience pays incredible dividends in terms of lives saved and victories won when units go into combat for real. (According to many returning Desert Storm veterans, combat in the Iraqi and Kuwaiti desert was just like it was at NTC, except that the Iraqis weren't as good sas OPFOR!) To put all this another way, failure in any area of the operational art will result in a failure of the unit in real combat, and so it is at the NTC.

Let's look at how Colonel Young and his troopers got ready for their 1993 NTC rotation.

Getting Ready

The 3rd ACR began to get ready for Fort Irwin in the late spring of 1993. The previous year, the regiment had sent only its 1st and 3rd Armored Cavalry Squadrons to NTC. Now Colonel Young would take the rest of the regiment to Fort Irwin, while 3rd Squadron deployed for an exercise in Kuwait (Operation Intrinsic Action 94-1). In order that the rest of the regiment would be better prepared for the coming NTC rotation, 3rd Squadron was used as an OPFOR unit for it to practice against. Since Fort Bliss butts up against the White Sands missile test range, the 3rd ACR enjoys the advantage of an enormous "back-yard" for maneuver and practice. In fact, the 3rd ACR has more room to maneuver than the whole III Corps facility at Fort Hood, Texas.

Meanwhile, since many of the Gulf War veterans had rotated out to other positions and most of the soldiers needed to requalify with their weapons, a program of live-fire gunnery "tables" was set up on a desert range north of Fort Bliss in June and July of 1993 (they are called "tables" because the scoring sheet is set up in rows and columns, with boxes for the evaluator to check off). Each table tests a different set of gunnery skills for a particular type of vehicle. It usually requires maneuvering up to firing positions, followed by a series of live-fire engagements with "pop-up" targets. Each vehicle crew must complete all twelve tables to be considered "qualified" in gunnery. The qualification process was long and dusty, with almost 200 3rd ACR crews requiring certification. When it was over, all the crews of 1st Squadron (Lieutenant Colonel Toby W. Martinez) and 2nd Squadron (Lieutenant Colonel Norman Greczyn) were fully qualified. These two units, along with the rest of the regiment (except for 3rd Squadron under Lieutenant Colonel Karl J. Gunzelman), were scheduled to move out prior to Labor Day weekend 1993 for the NTC rotation to Fort Irwin.

Even before the requalification, the two cavalry squadron commanders had begun teaching their units how to maneuver and fire as a group. First at the platoon, then at the troop level, Colonel Martinez and Colonel Greczyn worked their small units into teams that could maneuver at squadron level. This would be required of them at the NTC. It should be noted that this is not something special being done just for the NTC deployment, but the actual process that the regiment would go through if they were given a combat deployment order to go overseas. In fact, the entire regiment was involved in doing the same jobs they would have just before a combat deployment, from the support squadron getting equipment and supplies ready to ship, to the legal and medical sections updating wills and immunization shots.

After the requalification, squadron-level exercises were run (one squadron at a time) against an opposing unit made up of Colonel Gunzelman's 3rd Squadron, plus some Marines invited from the Twenty-Nine Palms desert training center. The unit would defend an objective on the range north of El Paso, with the troopers from either 1st or 2nd Squadron trying to dislodge them.

To keep score, there was a set of electronic gear called the Multiple Integrated Laser Exercise System (MILES). MILES provides a "no-shoot" way for units to practice ground combat (and some limited air combat), while scoring and recording their actions. MILES uses eye-safe lasers to simulate firing weapons; and MILES gear is available to simulate most weapons in the Army inventory, plus a few Soviet systems. Everything from main tank guns to assault rifles can be simulated with the right MILES transmitter.

In addition, small pyrotechnic charges (big, safe firecrackers) called Hoffman Devices are set off by the MILES electronics to simulate the noise and smoke when a gun or missile fires. Whenever a target is "hit" by the laser beam of a firing unit, a sensor on the target vehicle detects the laser "hit" and

A U.S. Army soldier with a full set of infantry MILES gear. The "buttons" on his web harness and helmet sense laser "hits" from enemy weapons, while his M16 assault rifle is equipped with a laser generator of its own.
OFFICIAL U.S. ARMY PHOTO

scores it either as a near miss or a simulated "kill." If a target vehicle is "killed," a yellow strobe light on top of the vehicle begins to blink, so everyone knows that it is out of action. To prevent arguments like "We shot you first," when a vehicle is "killed," the MILES system immediately disarms all of its lasers.

There are MILES systems for dismounted soldiers also, and they fit over the helmets and web gear of the users to score simulated personnel casualties. When a soldier in a MILES ensemble is "hit," the sensor harness lets out an unpleasant beeping sound for a "near miss" and a shrill continuous whine for a "kill," until the soldier disables it (and his weapon) with a small yellow key. In addition to "kills" scored by enemy vehicles, the "Observer/Controller" (O/C) referees (if present) can also simulate a kill with a small laser transmitter called a "God Gun" by the troops.

The fights begin with a night road march to what is called the "line of departure," the point where the attack actually kicks off. Along the line of march, "phase lines" are overlaid on the command maps. These imaginary lines, perpendicular to the line of advance, are usually given code names for clarity over the radio channels. First Squadron likes to designate phase lines by women's names, such as Debbie, Ginger, Zelda, etc. As each unit crosses a particular phase line, it radios a report to the squadron commander. This tells the commander if the attack is running on schedule, as well as who is doing what.

Frequently, the exercise director will throw in difficulties, such as simulated chemical and artillery attacks. For example, if a unit moves through a zone contaminated with persistent chemical agents (a referee would notify them of this), it must immediately don MOPP-IV chemical-protection suits, use its Fox NBC vehicles to survey the contaminated zone, report the results to squadron HQ, conduct decontamination if necessary, and then continue on to the objective.

In the event, these early, full-sized squadron exercises showed us the reason why soldiers train. A lot of things didn't go right. For example, right

A 3rd ACR M1A1 HC tank kicks up dust during exercises at Fort Bliss, Texas. *JOHN D. GRESHAM*

after the 4th of July break, we watched an exercise involving Colonel Martinez's 1st Squadron. The objective was a hilltop in the eastern part of the Fort Bliss range overlooking 1st Squadron's approach from the southwest. To make things harder, 3rd Squadron and the Marines had been given several days to dig in on the objective and along the approach route.

During the pre-exercise briefing the day before, we tried to motivate the colonel and his officers by promising "donuts and coffee" at the objective—but our good intentions didn't help them much. 1st Squadron became a victim of what Clausewitz (the great German philosopher of war, 1780–1831) called "friction": One little thing after another conspires to keep you from achieving your objective. "Everything is very simple in war," Clausewitz wrote, "but the simplest thing is difficult. These difficulties accumulate and produce a friction which no man can imagine who has not seen war."

In this case, dust and the cover of darkness caused a few platoons to lose their way and become separated (even with GPS, this sometimes still happens). Meanwhile, some Marine OPFOR teams slipped into 1st Squadron's rear assembly areas and caused additional confusion. By the time 1st Squadron sorted itself out for the attack on the hill, it was being chopped up by 3rd Squadron's dug-in tanks and Bradleys.

Now, you might ask:

Q: What is the point of an attack exercise that fails?

A: You learn from your mistakes.

A formal Army review process makes sure that participants learn well. This "After-Action Review" (AAR) is a post-exercise meeting, with all leaders from both sides describing what they saw. The referees go over every detail of the exercise in excruciating detail. And it is expected that the commander of each unit will stand up and give a candid self-criticism of just what happened. If he is unsure of what went wrong, the O/Cs may ask the opposing commander to explain what happened. To further clarify all this, MILES data is displayed to show the movements and lines of fire. When it is all over, "lessons learned"

The staff of the OPFOR at a pre-mission brief. The author and series artist Laura Alpher are sitting in the front row of seats.
JOHN D. GRESHAM

are written down for later distribution to the participants. No Catholic confessional could ever be so uncomfortable and candid as a properly run After-Action Review.

By the way, we gave the troopers of the Tiger Squadron the donuts anyway!

The Move to the NTC

When the squadron exercises were completed, the regiment was ready for the move to Fort Irwin. Several weeks before, Colonel Young and his officers visited the NTC to be briefed in great detail on rules of engagement (ROE), range safety procedures, logistics instructions, and other procedures. No effort is spared to ensure that the deployment is a successful learning experience. It was to be a learning experience for me as well; I spent several days observing the 3rd ACR's NTC exercises.

In order to save wear and tear on its equipment, when a unit goes to NTC, it usually draws most of its vehicles and equipment from a storage depot at Fort Irwin. Consequently, only a few 3rd ACR command vehicles had to be shipped there.

Just prior to Labor Day weekend, most of the soldiers were trucked and bused to Fort Irwin (about 10% went by air) to draw their equipment and head out to the range assembly area called "the dust bowl."

At the same time, 4th Squadron flew their helicopters directly to the Fort Irwin airfield, and then moved out to a field base where they would operate during the exercise. The regiment now split into two parts for the duration of the exercise. For the first half of the exercise, 2nd Squadron (Lieutenant Colonel Greczyn) went up to the live-fire range. The rest of the regiment headed out to the maneuver area in the southern part of Fort Irwin for their series of "fights" with the toughest unit to *never* serve in the Soviet Army, the OPFOR.

The National Training Center Facilities/Staff

When you drive east on Interstate 10 from Los Angeles, you cross Interstate 15, which heads north towards Las Vegas, Nevada. About halfway to "Sin City" on I-15 lies the desert town of Barstow, gateway to the National Training Center, which lies a further thirty-seven miles north. The first thing you notice as you drive to the base are numerous white crosses—memorials to drivers killed along the road to NTC over the last dozen or so years. They are a grim reminder that life at NTC is dangerous enough without reckless or drunken driving. A more intriguing monument is the colorful pile of painted rocks decorating the base entrance. Each bears the crest of a unit that has rotated through.

Fort Irwin is desolate and treeless—just lots of rocks, dust, small brush, and open space for maneuvering and shooting. And yet, amazingly enough, the Army feels confined in its roughly 1,000 square miles/2,687 square kilometers.

So it plans to acquire another parcel of land to enlarge the facility by 50%. Currently limited to battalion/squadron-sized actions, the range, when expanded, will be able to host brigade/regimental-sized fights.

Though thousands of troops and their families live and work here, with more arriving all the time, only one combat unit, the 177th Armored (OPFOR) Brigade, is stationed on post. The post itself is not an elegant place. Most of the buildings were built in the no-frills style of government cinderblock architecture after NTC opened in 1981. Before then, the main activity at Fort Irwin was NASA's Goldstone Deep Space Tracking Facility, a huge dish antenna for communicating with space probes. As for the training facility, that is very elegant indeed. The entire range complex is instrumented with a huge MILES-based tracking and scoring system, and there is equipment meant to simulate combat with the highest possible fidelity.

During our visit in late 1993, Brigadier General Robert S. Coffey, a career infantry officer, commanded the NTC and the following units:

- **Fort Irwin Base Garrison**—The permanent force assigned to Fort Irwin runs everything from the Post Exchange to the base housing facilities. It provides maintenance and upkeep for the store of equipment used by units rotating through NTC.

- **The NTC Operations Group**—These soldiers take care of the MILES sensors, laser weapons, and target arrays. They also run the "Star Wars" building where NTC exercises are monitored and AARs are conducted. They supervise combat exercises on the range and also act as coaches, trainers, referees, range safety officers, data collectors, and target array operators.

- **177th Armored Brigade/60th Guards Motorized Rifle Division**— This is the maneuver element of the OPFOR. Based on Soviet organization and doctrine, they simulate enemy forces for the other units training at NTC. They operate old M551 Sheridan light tanks and HMMWVs with sheet-metal add-on kits ("visual modifications" or VISMODS) that make them look like Soviet-designed vehicles. With the apparent end of the Soviet threat, the OPFOR is keeping up-to-date on the tactics and organization of opponents America might face in the Balkans, the Middle East, East Asia, and other trouble spots.

These groups work together to make the NTC the most comprehensive combat training facility in the world. Virtually everything that a combat unit can encounter in a field deployment is simulated at the NTC, and units often find that a trip to Fort Irwin is tougher than actual combat.

Force-on-Force Fight—Wednesday, September 8th, 1993

When Colonel Young, Lieutenant Colonel Martinez, and Lieutenant Colonel Sealock arrived at the NTC, their first task was to get both 1st and 4th

A UH-1 helicopter assigned to the OPFOR at the NTC banks after a simulated firing run on troops of the 3rd ACR. The VIS-MOD on this helicopter allows it to simulate a Soviet/Russian HIND-D attack helicopter.
JOHN D. GRESHAM

Squadrons and the regimental HQ settled in and ready for action in the series of force-on-force exercises scheduled for the first ten days of the three-week rotation. These simulated battles put the squadrons through a series of engagements with the OPFOR. Every other day for eight days, over 5,000 soldiers would meet in battle up on the range. The scenarios are devised by the O/C group, and their aim is to confront participants with a variety of operational challenges. These can take several forms. Sometimes the two forces are moving simultaneously, and wind up fighting over some piece of desert, wherever their patrols collide. Other times the OPFOR attacks the unit (called the Blue Force) to break through to seize a set objective. Or the reverse: Blue Force might attack the OPFOR, in order to try to seize an objective.

Whatever the situation, both sides plan their actions so that the training experience for the Blue Force will be maximized.

Now, from the viewpoint of the OPFOR, maximizing the training experience means doing everything possible to beat a visiting Blue Force, up to and including simulated weapons of mass destruction. And in fact, the OPFOR wins something like 80% of the battles against Blue Force opponents. For the odds are intentionally stacked in the OPFOR's favor, making every battle a desperate fight for life by the Blue Force. Thus any fight that Blue "wins" requires perfect execution of their own battle plan, and usually some mistakes by the OPFOR. Anything less than perfection means annihilating defeat for the Blue Force, and an AAR with enough humbling self-criticism to fuel a lifetime memory—and a lifetime lesson. Some factors that account for the OPFOR's track record include:

- **Scenario Force Ratios**—The force ratios for each scenario (depending on Blue Force maintenance and readiness) are determined by the Operations Group, so that the fights are as tough as possible. Since U.S. units are expected to fight and win against enemy forces of larger size, the OPFOR units are up to twice as large as their Blue Force opponents, whether the OPFOR is attacking or defending.

Col. Bob Young (front) and the command staff of the 3rd ACR's 4th (Air Cavalry) Squadron plan an NTC force-on-force engagement on a "sand-table" model of the NTC range.   *John D. Gresham*

- **Weapons**–The NTC makes no differentiation between the effectiveness of American weapons (simulated by the MILES gear) and those of the "Soviet-equipped" OPFOR unit. This means that OPFOR tanks and fighting vehicles get the same performance from their guns and missiles as the U.S. units.

- **Home Field Advantage**–During any six-week period, the OPFOR fights between eight and ten battles on the same terrain, frequently in the same tactical situations. This means that they "fight" more often than any unit in the Army. Like an inner-city high school basketball team playing on its own court, they know every loose floorboard and rough spot by name. Frequently, their Blue Force opponents are seeing the NTC for the first time–which means they pay dearly while adapting to unfamiliar ground. OPFOR units use terrain so skillfully that first-time NTC visitors joke about the "OPFOR Tunnel," as if the OPFOR could "pop up" out of the ground anywhere it chooses.

- **OPFOR Experience**–Every member of the OPFOR is an experienced armor officer or enlisted soldier. They all participate in the OPFOR mission constantly, and that means that they are good at their jobs. As in life, so in the combat arena, practice makes perfect.

For several reasons, the situation for 3rd ACR was even tougher on this rotation. For starters, because cavalry troopers consider themselves to be an elite, the OPFOR seems to have a special place in its heart for the cavalry. Therefore, whenever the 2nd ACR-L or the 3rd ACR come to the NTC, the OPFOR likes to knock them down a couple of pegs by devising extra-tough scenarios and taking longer risks to win.

In the event, the first few days of force-on-force battle did not go well for the 3rd ACR. Though they got close to winning once, they were beaten particularly badly on several occasions. So, when the warning order came in on the morning of September 7th for a movement-to-contact the next day

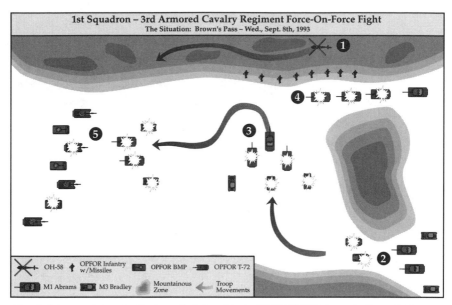

1st Squadron – 3rd Armored Cavalry Regiment Force-On-Force Fight
The Situation: Brown's Pass – Wed., Sept. 8th, 1993

Legend:
OH-58 | OPFOR Infantry w/Missiles | OPFOR BMP | OPFOR T-72
M1 Abrams | M3 Bradley | Mountainous Zone | Troop Movements

The 3rd Armored Cavalry Regiment's fight with the NTC OPFOR forces in Brown's Pass on September 8th, 1993. As the fight developed, (1) the helicopters of 4th Squadron scouted the north wall of the pass with only limited success. Then (2) 1st Squadron destroyed the OPFOR security force in the southern entrance of the pass, then moved up to the OPFOR positions along the north wall. Here (3), they suffer from OPFOR missile fire. The 1st Squadron's tank company also is hit heavily, and the survivors are destroyed (5) when they run into the main OPFOR force.

Jack Ryan Enterprises, Ltd., by Laura Alpher

(everything down to the operations orders are handled as if they were actual combat situations), Lieutenant Colonel Martinez knew that it was the last chance for a successful fight with the OPFOR. Following this exercise, they would move north to the Drinkwater Lake Live-Fire Range for their turn at the NTC's shooting gallery.

September 7th was spent repairing equipment (armored vehicles wear out at a vicious pace in the desert) and planning the coming battle. Meanwhile, at the 1st and 4th Squadron HQs, the commanders drew battle plans on large sand-table models of Fort Irwin. The objective was to push through Brown's Pass on the northern end of the range. The staff of the 3rd ACR expected to find the OPFOR dug in on the far side of the pass, though exactly where depended on the pre-battle reconnaissance scheduled for that evening. If they could just find the OPFOR, they would blow them right out of Brown's Pass and back to the Goldstone tracking center.

At a "lager" (an encampment of combat vehicles, typically in a circle facing outward—think of covered wagons) near Goldstone, the leaders of the OPFOR were also planning for tomorrow's battle. After several hours of work by the operations (S-3) staff, it was decided that the battle would begin with aggressive counter-reconnaissance against the 3rd ACR to the east (whoever wins the reconnaissance battle usually wins the fight). Vehicles of the 60th Guards Motorized Rifle Division would dig in on hills north of the pass. Anti-

Lieutenant Colonel Toby Martinez, the commander of the 3rd ACR's 1st "Tiger" Squadron, after the defeat in Brown's Pass. He had "jumped" from four vehicles that had been "hit" during the force-on-force engagement. *JOHN D. GRESHAM*

Tank Guided Missile (ATGM) teams would hide among the rocks at the mouth of the pass. When the 1st Squadron came through the next morning, they would enter an ambush with interlocking fields of fire.

As both sides moved up to their initial positions that night, a series of little actions began that would have a decisive impact on the following day's battle. Along the line of hills and ridges north of the pass, the reconnaissance teams that 3rd ACR landed from 4th Squadron's Blackhawk helicopters were quickly snatched up by counter-reconnaissance teams from the 60th Guards. At dawn the next morning, as 3rd ACR's 1st Squadron moved west up into the pass, they were relatively blind.

Their only reconnaissance came from 4th Squadron's aero scouts, who reported an enemy security outpost in the pass itself, and enemy armored vehicles in fighting positions to the west of the pass. But they could not get close enough to see more, due to the threat of simulated SAM and anti-aircraft fire from OPFOR positions.

At dawn the next morning, 1st Squadron blasted right through the pass, losing only a few vehicles to the OPFOR troops there. They then made a long bounding movement to the northwest towards the dug-in armored vehicles reported by the aero scouts. Lieutenant Colonel Martinez rode into battle in his M3 Bradley. Just behind came Colonel Young's regimental command section, a pair of Bradleys.

During the bound towards the northwest, M1059s (M113 APCs with smoke generators) ran along the flanks of the lead vehicles laying down smoke, and all seemed to be going well. But suddenly all hell broke loose, as the hills to the north of the pass erupted in a hail of simulated ATGMs. A number of Abrams tanks and Bradleys were "killed" in this first barrage, and the plain to the west of the pass became littered with unmoving vehicles topped by blinking yellow strobe lights. Worst of all, the command tracks of Colonel Young and Lieutenant Colonel Martinez were knocked out in the missile barrage.

Toby Martinez is not a soldier who gives up easily. As soon as his command track was hit, he "jumped" to another vehicle to keep the momentum of the attack rolling. He would do this three more times before the day's battle ended. Back on the command radio net, he ordered the survivors of A, B, and C Troops to attack and clear out the OPFOR missile teams from the rocks in the hills. While they were doing this, he sent D Company, the reserve of fourteen tanks, to continue the attack towards the dug-in armored vehicles to the west.

Unfortunately for Lieutenant Colonel Martinez and the troopers of 1st Squadron, the failed reconnaissance of the previous evening returned to haunt them. When the tanks of D Company reached the line of dug-in T-72 tanks and BMP fighting vehicles, they found not the expected company-sized unit, but almost a whole battalion of OPFOR fighting vehicles. Split into two widely separate groups, the squadron's units were defeated in detail before they realized what had happened. The ATGM teams had pulled the cavalry troops in like flypaper, and the iron fist of OPFOR crushed them against the rock walls of the pass. Less than four hours after 1st Squadron left the line of departure, it had been decimated. Around the western mouth of the pass, the blinking "killed" lights of 3rd ACR vehicles bore mute testimony to the prowess of the OPFOR.

Nobody in the regiment had a tougher break that day than Toby Martinez. Not only did he have to "jump" four times from vehicles that were shot out from under him (he was looking for a fifth when orders terminating the battle came in over the radio net); but while he was jumping down from the fourth vehicle, an Abrams tank, he pulled muscles in his side and back and injured his right knee. This mishap would leave him in severe pain for days— still commanding his squadron. For now, though, he and Colonel Young rushed back to the After-Action Review (AAR) at the "Star Wars" building to take their knocks.

The lessons learned at this particular AAR included suggestions on:

- Improving the regiment's reconnaissance and counter-reconnaissance plan.
- Better use of artillery and close air-support missions, which had almost no effect on the action in the pass.
- Better use of the terrain in the pass to avoid exposure to ATGMs during long transits in the open.

With this fight over, their fourth in just a week, 1st Squadron made preparations to switch with Norm Greczyn's 2nd Squadron up at Drinkwater Lake. This would involve a 30-mile/50-kilometer road march by all of the squadron's vehicles to the other end of Fort Irwin. During this time, Colonel Young and

The author with General Coffey (left) and series illustrator Laura Alpher (center) at a field meeting of NTC observer-controllers after an exercise. *JOHN D. GRESHAM*

the regimental tactical operations center (RTOC) team would stay behind to work with 2nd Squadron for another ten days of force-on-force battles. It had been a tough week for the troopers of the 1st Squadron, and they were already tired, dusty, and dirty. The previous few days had been extremely hot and it would get hotter as they moved east to the live-fire range. Temperatures topped 115°F/46° C. But inside a tank or Bradley, the temperature can get up to 20°F /11°C *hotter*! For myself, I must confess that the heat was devastating. I was still getting into the "groove" of the hydration cycle, and more than once I wished I had stayed back in the air-conditioned luxury of the VIP quarters at Fort Irwin. As we flew back to the main base in a UH-1 helicopter for a "dining-in" that evening, I kept thinking about the young 3rd ACR troopers who would not have a fresh meal, a shower, or an air-conditioned room to sleep in. At the same time, I found myself wishing better luck to Toby and his squadron in their first live-fire at Drinkwater Lake on Sunday morning.

Daytime Live-Fire, Sunday, September 12th, 1993

If Lieutenant Colonel Martinez had hoped for better luck following the 1st Squadron's force-on-force phase of the NTC deployment, then he had some cruel surprises coming his way. His injuries from the fall on the 8th were still hurting badly, and he was getting only a couple of hours sleep a night. Sleep deprivation is a serious matter when you are making life-and-death decisions, and the pain in his side did not help. Worse yet was the news about the availability of tanks and Bradleys. By Sunday morning, due to shortages of spare parts at Fort Irwin, only 60% of his tanks and 55% of his Bradley fighting vehicles were operable. This meant that engineer support and indirect artillery fire would play a much larger role than usual. It would be the deciding factor in the success or failure of the squadron.

What the squadron faced that Sunday morning is the mother of all shooting galleries: Drinkwater Lake Live-Fire Range. Nestled in a valley in the northeast corner of Fort Irwin, it runs east to west, with hills on the north side and rugged mountains to the south. At the eastern end of the valley, at the bottom of a long slope, is Drinkwater Dry Lake, a seasonal lake dried smooth and white by the vicious summer heat. On the floor of the valley are a series of 1,500 computer-controlled "pop-up" targets that simulate visually the advance of an MRR down the valley towards the positions of the Blue Force.

The targets are controlled by the O/C team at the range control center perched atop a hill on the western end of the valley. When a target is hit by one of the training-practice (TP) rounds fired from a tank or fighting vehicle, it is scored as a "kill" and does not appear again in the line of "pop-ups." To win, the Blue Force has to destroy three full battalions of targets, or about 160 simulated enemy vehicles. The conditions at Drinkwater Lake are as real as range safety regulations will allow. Live artillery is allowed (as we would see later), and even chemical warfare is simulated, with the use of tear gas grenades requiring the troopers to fight in MOPP-IV suits.

The day prior to the live-fire exercise, Lieutenant Colonel Martinez, his commanders, the S2, the S3, the fire support officer, the direct support field artillery battalion commander, and the engineer company commander walked the battlefield and decided on the engagement area where they intended to kill the enemy. After the leaders' reconnaissance of the terrain, the squadron commander directed the positioning of the vehicles, identified the limits of the engagement area, and divided it into three separate kill zones. The first zone, out of range of direct-fire weapons, was dominated by the artillery and the close air support (CAS) F-16s. The intent was to mass on the enemy the indirect fires of the regiment's field artillery battalion (consisting of three artillery batteries) and the squadron's own howitzer battery (thirty M109 tubes total). Lieutenant Colonel Martinez made it clear he wanted all artillery batteries massed on enemy formations using "linear sheafs"[1] for maximum destruction. If the enemy was allowed to enter within the range of direct-fire weapons unharmed, there would be too many targets for the tanks and Bradleys to kill and the enemy would roll over the positions. The fire support officer, Captain Joe Feistritzer, positioned the fire support teams (FISTs) on the high ground to the flanks of the engagement area, to ensure long-range and overlapping fields of observation. Artillery targets were identified and surveyed, and trigger points were marked on the ground. The second kill zone was dominated by the M1 Abrams and TOW missiles from the M3 Bradleys, based upon the maximum range of these weapons. The priority targets for this kill zone were the enemy tanks. The intent was to destroy the enemy tanks before they could close within their effective fire range. Artillery and mortars would continue to provide suppressive fire and smoke. The squadron commander expected to destroy the remaining enemy within the third kill zone. This is where the majority

(1) 3rd ACR M1A1 HC tanks (the one on the left is equipped with a mine plow) after a force-on-force engagement in Brown's Pass. *John D. Gresham*

(2) 3rd ACR troopers in chemical-warfare suits work around their M109 self-propelled howitzers. *John D. Gresham*

(3) A 3rd ACR M1A1 HC tank dug in on the Drinkwater Lake Live-Fire Range at the NTC. *John D. Gresham*

of the obstacles were emplaced, and where all weapons systems could reach the enemy. The squadron commander knew that if the artillery could kill at least 20% of the targets at long range, and if the engineer obstacle plan could keep the enemy in the engagement area for at least ten extra minutes, the troopers manning the tanks and Bradleys would finish the destruction of the enemy.

Once the plan was set, all components of the squadron were put in motion toward the same objective—preparing the battlefield to destroy the enemy. Late that night, Lieutenant Colonel Martinez met with his commanders—the artillery battalion commander, Lieutenant Colonel Twohig; the Tiger Squadron operations officer (S3), Major Brossart; the fire support officer, Captain Feistritzer; and the squadron intelligence officer (S2), Captain Whatmough—and one more time went over in detail the engagement area, fire control, engagement priorities, every artillery target trigger point, and the synchronization matrix. Following the meeting, he met with his squadron executive officer, Major Sandridge, to get the latest combat power report; and then he boarded his M3 Command Track with his fire support officer and moved to his position for the morning's fight. The following morning at "Stand To" (the designated time in which all weapon systems and personnel must be ready to fight), he received status reports from all commanders, and an intelligence update from the S2. At this point, Lieutenant Colonel Martinez realized there was nothing else he could do to prepare the Tiger Squadron for battle. He tried to relax and waited for the enemy.

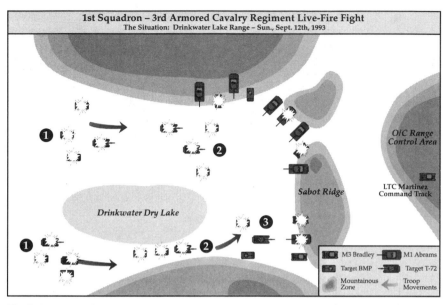

1st Squadron – 3rd Armored Cavalry Regiment Live-Fire Fight
The Situation: Drinkwater Lake Range – Sun., Sept. 12th, 1993

O/C Range
Control Area

Sabot Ridge

LTC Martinez
Command Track

Drinkwater Dry Lake

M3 Bradley — M1 Abrams

Target BMP — Target T-72

Mountainous
Zone

Troop
Movements

The 3rd Armored Cavalry Regiment's fight with the NTC's live-fire target array on September 12th, 1993. The fight developed (1) with the first wave of two simulated motor rifle battalions (MRBs) being hit by well-planned and directed artillery fire. Then (2) 1st Squadron destroyed the rest of the two MRBs with fire from positions on the north side of the valley and Sabot Ridge. The last assault (3) came from the reserve MRB along the south side of Drinkwater Lake. Due to a stop in the artillery fire for safety reasons, a few (about 25 out of 160) simulated vehicles got through. *Jack Ryan Enterprises, Ltd., by Laura Alpher*

With less than 60% of his direct-fire weapons available that morning, the squadron would have to come up with something special to just break even in this battle. As we chatted on the flight out, General Taylor (himself a former NTC commander) told me that with their vehicle availability problems, he had little hope of 1st Squadron doing very well that day, and that he was looking forward to seeing "how much character Toby and his troopers have." All of us were going to get an education on just how much they did have.

By 0700 (7:00 AM), we had landed and were having coffee with the O/C team at an observation area overlooking the valley. The O/C team's plan for defeating the crippled 1st Squadron was brutal and simple. The target controllers would run one battalion of targets along each side of the lake from east to west. And they would commit their "reserve battalion" to whichever thrust got the farthest. This is consistent with the Soviet-style doctrine of reinforcing success with combat reserves. As the first line of targets "popped up" on the saddle of the pass leading into the western end of the valley (called the "S-Band" of targets), the O/Cs were confident that they would blow 1st Squadron right out of the valley.

With the positions of the defending Blue Force vehicles fixed (due to range safety requirements), they could not retreat or shift firing positions during the battle. This made the setup of Colonel Martinez's already depleted force even more critical. The main line of defense ran from the base of the

Dug in to a shallow fighting position at the NTC, this Bradley prepares for battle with the OPFOR. Note the TOW launcher in its ready-to-fire position, and the blinker lamp mounted beside the commander's hatch.

mountain on the southern side of the valley, along a raised piece of ground called Sabot Ridge, and then in a long crescent along the north side of the valley. The setup was designed to provide maximum interlocking fire against the widest corridor in the valley along the north side of the dry lake bed. Because the brine shrimp larvae that hibernate beneath the surface are an endangered species (no, I am *not* kidding), one of the rules of the live-fire range is that nobody can intrude upon the dry lake, and no heavy vehicle traffic or artillery is allowed.

Meanwhile, the O/Cs had us don flak jackets and helmets for protection in case a stray round should fall nearby. And at 0800 (8:00 AM), we sat down to watch the battle. It was like a slow pageant, with the first line of targets appearing and advancing silently. Occasionally, the O/Cs would have a target "pop" a smoke generator to simulate the dust of vehicles moving across the desert floor. Just then, the first ranging shots from 1st Squadron's artillery batteries began to strike around the advancing line of targets. Immediately, the artillery began to "kill" large numbers of targets, and the northern target line began to crumble. The noise of the artillery was like distant thunder, with the black smoke and dust of the HE shells temporarily obscuring our view. Then, the first of the F-16s from Nellis AFB (near Las Vegas, Nevada) came in, hitting even more of the northern targets. To the south, the same thing was happening, with the artillery chewing up the advancing target line along the southern edge of Drinkwater Lake.

Lieutenant Colonel Martinez's artillery and CAS plan worked so well that we almost forgot about the depleted force of tanks and Bradleys in the valley waiting for "leakers" that got through the barrage. Suddenly, the first tank opened up with a "crack" from its 120mm main gun. Almost immediately, several others fired; and we heard the "whoosh" of TOW missiles heading out across the lake bed to hit targets trying to move around to the south. As the simulated enemy advance reached the line of obstacles emplaced by the engineers, the last weapons to engage were the 25mm chain guns on the Bradleys. The fire discipline was perfect, with crews firing single ranging shots followed

The commander of III Corps, General "Pete" Taylor, talks with Lieutenant Colonel Toby Martinez, the commander of the 3rd ACR's 1st "Tiger" Squadron, after their NTC live-fire fight at Drinkwater Lake.

JOHN D. GRESHAM

by three-round bursts when they found the range. Over the O/C radio net, we could hear the count of targets destroyed, and were surprised when the controllers announced that the first two battalions had been annihilated, and that they were going to commit the reserve battalion of targets to the southern corridor around the lake bed.

There was a temporary lull as the O/Cs moved up the last of the battalion target arrays and the artillery crews got ready to re-engage with their big guns. When the M109s started firing again, they repeated for a time the success of the previous barrage. Even the dropping of some tear gas grenades around the defense line (causing them to don their MOPP-IV suits) and the liberal use of "God Guns" to "kill" random tanks and fighting vehicles did not slow down the 1st Squadron fire.

Then the inevitable "friction" of war came into play. Several 155mm HE rounds from an M109 battery impacted at the wrong location, throwing up a shower of salt and brown earth. NTC safety rules require the O/Cs to issue a "Cease Fire" order to all batteries when this happens, so the artillery chain of command can sort out who is shooting at what. It only took ten minutes or so to sort it out and get the guns back firing. But in that time, the target array had moved around the lake and into the eastern end of the valley. In the end, when the order arrived terminating the battle, a handful of "leakers" had reached the defense line along Sabot Ridge. As the guns stopped firing and the "All clear" signals came in over the net, you could hear the usually calm and objective O/C team's excitement over how well 1st Squadron had done. Clearly, we had witnessed something special at the NTC: A unit had snatched "victory" from a seemingly hopeless situation. When the final tally came in, Toby and his troopers had "killed" 135 out of 160 simulated enemy tanks. Although they had suffered some twenty-five "leakers" at the end, they had racked up one for the NTC record book.

Doffing our body armor and hopping into a couple of HMMWVs, Generals Taylor, Coffey, and my team headed up the hill to the 1st Squadron command team to meet with Lieutenant Colonel Martinez and find out what he

Medical personnel of the 3rd ACR's support squadron treat simulated medical casualties after an NTC live-fire exercise at Drinkwater Lake.
JOHN D. GRESHAM

had seen. As we drove up to the pair of M3 command tracks at the top of a ridge, Toby Martinez was pulling himself out of the small rear hatch to greet us. You could clearly see that he was in pain just standing there breathing. But through the pain you could also see a profound exaltation at what his troopers had done. As he briefed General Taylor on how the fight had gone, he assured Pete that he would not only get more vehicles back on-line for that evening's live-fire shoot, but that they would "kill every damned one of those targets tonight, sir!"

Clearly excited themselves, Generals Taylor and Coffey made sure that Toby and his TOC staff knew just how well they had done that morning. In just two hours (it seemed like about twenty minutes with things happening so fast), the squadron's attitude had been turned around by the amazing performance of Lieutenant Colonel Martinez's preplanned artillery fire. Now NTC "winners," they could look forward to other engagements with the knowledge of their "good fight." It was a rare privilege to watch Pete Taylor, a veteran of shooting wars with three stars, put his arm around the young lieutenant colonel and tell him, "Someday you'll have a regiment of your own, son."

Aftermath

Happy as Toby was over the outcome of the battle, he and the squadron had things to do before they could grab a few hours of sleep. The squadron had to refine the artillery observer positions and artillery targets, artillery batteries needed to reposition, barriers had to be reseeded, simulated casualties had to be evacuated properly, and repaired vehicles had to be brought forward from the Unit Maintenance Collection Point (UMCP, the slang term is "Bone Yard"). Failure to do all that would cause the O/C team to downgrade the squadron's performance, and so undo their success. At the same time, ammunition and food had to be brought forward and vehicles reloaded, and if necessary (or possible) repaired.

The early performance had been amazing. But it got better when the sun set later that day. Getting a few more tanks and Bradleys out of the repair yard

The commander of III Corps, General "Pete" Taylor, talks with Colonel Bob Young, the commander of the 3rd ACR, after a force-on-force engagement at the NTC.   *John D. Gresham*

and shifting more forces toward Sabot Ridge in the eastern end of the valley, they killed even more targets that evening, demonstrating the spirit of the 3rd ACR troopers under adversity. In that moment of victory on Sabot Ridge on Sunday morning, they reaffirmed the regimental motto, "Brave Rifles!" and its response, "Blood and Steel."

Back over in the force-on-force area, though, 2nd Squadron was having their own troubles with the OPFOR. The previous day, fresh from several excellent live-fire performances of their own, they had also suffered heavily up near Brown's Pass. In an engagement with a superior OPFOR force, the scouting broke down again, Colonel Young had difficulties getting information over the radio nets to decide when and where to commit the squadron's reserve company of tanks. The OPFOR blasted through the northern part of their over-extended line; and in just seventeen minutes, OPFOR tanks overran the regimental supply and trains area. It was a tough learning experience for the new regimental commander, who was going through his first NTC rotation. But learning is what NTC is all about, and Bob Young took his knocks like a professional.

The regiment had another week of simulated combat, producing a smattering of successes, as well as the inevitable losses to the OPFOR. As they finished up the rotation in the third week of September, they turned in their borrowed vehicles and equipment, and boarded the charter buses back to El Paso. A couple of weeks later, the NTC staff shipped them the payoff for all the efforts—the 1,300-page "take-home" package. This is a catalog of every move and shot by the regiment during their entire rotation. This package provides guidance on what the regiment needs to work on until they come back for their next NTC rotation in the summer of 1994.

While most of the regiment was at the NTC rotation, back home at Fort Bliss, Lieutenant Colonel Gunzelman and his 3rd Squadron were getting ready for deployment to Kuwait later in the fall. At the end of 1993, 3rd Squadron finished up Intrinsic Action 94-1 and came home for the holidays. The regiment was together for only a short time, however, before the next set of training rotations.

Because of all this hard work and dedication, the 3rd ACR will remain one of the units that the Army can count upon to be "ready" in case of an emergency or crisis. And the troopers of the Brave Rifles will continue their almost 150-year tradition of service to the United States of America.

A Cavalry Officer's Life

In previous chapters, we have explored the equipment and institutions that allow the U.S. Army to train and equip soldiers and make them into a working part of a cavalry or armored unit. But what about the human part of all this? Just what does it feel like to join the Army and dedicate your professional life to it? Our talk with General Franks provides some telling insights, of course, but what does the Army life mean for someone younger, less senior? How does a young man or woman just starting out see it, in those early years of young adulthood when the world is just opening up and all of life is a new adventure?

To get a feel for how it is for a young person to live in today's Army, let's get to know one of the service's brightest and most successful young officers. Herbert Raymond McMaster, Jr.—H.R. to his friends—was born on July 24th, 1962, in Philadelphia, Pennsylvania. A West Point graduate in 1984, H.R. has gone on to a rewarding career in the Army. As commander of Eagle Troop of the 2nd Squadron of the 2nd Armored Cavalry Regiment during Operation Desert Storm, he led a highly successful attack on a brigade of a Republican Guard Division. This attack—in what has come to be called the Battle of 73 Easting (a reference to a map grid position in Iraq)—is generally considered to be a textbook example of small-unit operational art and command initiative. Not despite this impressive accomplishment, but because of it, H.R. is representative of the new generation of soldier that has joined the Army in the last twelve to fifteen years. Smart, fit, attractive, and committed to the ideals that made him join the service. Let's meet him.

Tom Clancy: Please explain what your childhood was like, and in particular, what was it that made you decide to go into the military?

H. R. McMaster: My father served as an infantryman during the Korean War and joined the reserves as a first sergeant in his early twenties. He was awarded a direct commission from first sergeant to captain during the Vietnam War. He later went on active duty and secured assignments in the greater Philadelphia area. I have one sister, Letitia, who is two years younger than I am. She is a graduate of Villanova University, and now works for a computer firm and is pursuing her MBA. My mom is a recently retired school teacher of thirty-seven years or so. She began teaching fourth grade and then focused on kindergarten. She earned an MED degree and became a school administrator. I consider her to be a brilliant educator. In retrospect, I think that the professional satisfaction she gained from having a positive effect on people lives, as well as my father's Army experience, made service in the military appealing to me. I attended high school at Valley Forge Military Academy, which is about ten miles from my home in the Roxborough section of Philadelphia. It was a great experience in terms of learning about leadership at a very young age. In both grade school and high school, I participated avidly in team sports, football and baseball in particular. I think that what one learns from playing team sports, particularly about how to work with others toward a common goal, carries over directly to what is required to serve successfully in the military.

Tom Clancy: Tell us about your college decision.

H. R. McMaster: During my senior year at Valley Forge, I applied both for an appointment to the United States Military Academy at West Point and an ROTC scholarship at the University of Notre Dame. I had a very difficult decision to make between these two fine institutions. I based my decision, in part, on the belief that West Point would prepare me best for a military career. The unique challenges of West Point also intrigued me. So, in July of 1980, I entered as a new cadet. The transformation from civilian to cadet is accomplished quite rapidly. In fact, by the end of the first day, the new cadets put on a parade for the parents, who watch from the reviewing stands. During this review, the entering class takes the oath of service to the United States.

Tom Clancy: So, in other words, by the end of your first day at West Point, you were already being recognized as a member of the United States Armed Forces.

H. R. McMaster: Yes.

Since the early 1800s, Americans have sent their sons—and more recently, their daughters—to join the "long gray line" of West Point's Corps of Cadets.

Through the post gates overlooking the Hudson River have passed some of the greatest officers in military history. Names like Lee and Grant. MacArthur and Eisenhower. Bradley and Patton. Schwarzkopf and Franks. The Academy positively oozes history and tradition, and H.R.'s early memories of the post tell us a great deal about such things.

Tom Clancy: What did you think of your first summer at West Point?

H. R. McMaster: West Point is imposing architecturally. And the campus exudes tradition and military professionalism. A new cadet is filled with excitement and pride at becoming part of the tradition—a member of "the long gray line." You learn rather quickly that the key to success in "Beast Barracks" [the slang term for the pressure of barracks life during the first summer] is teamwork and selflessness. The pressure that upperclassmen put on new cadets drives home the point that all must work together to get through this initial, intense period. This is an important lesson for the aspiring officer. New cadets are assigned innumerable tasks to accomplish in an inadequate amount of time. And one quickly learns how to prioritize tasks and accomplish the most important ones first. Because no one cadet can accomplish all of this, he or she can expect a certain amount of negative attention. As you can imagine, it's important to keep one's sense of humor!

Tom Clancy: What were the barracks like, and who were the people you lived with?

H. R. McMaster: I was first assigned to a three-man barracks room. There I met Dan Miller, and we quickly became great friends. By coincidence, we also ended up in the same company at West Point after the shuffling that followed plebe [or freshman] year. And we met again six years after graduation in the 2nd Armored Cavalry Regiment [2nd ACR], where he commanded Iron Troop in the 3rd Squadron and I commanded Eagle Troop in the 2nd Squadron. My other roommate had a difficult time adjusting to West Point. Dan and I tried to help him through, but he just couldn't cope with the pressure. One day, as Dan and I were leaving for training, he told us that he was quitting. When we returned, he was gone. He decided that the military was not the right profession for him, and he withdrew from the Academy.

Tom Clancy:	How do you feel about your roommate's decision? Do you respect him for it?
H. R. McMaster:	I respect his decision without qualification. He and many other good friends departed the Academy at various points during my four years there. In fact, if you asked my own tactical officer, he probably would tell you that I also came precariously close to leaving—not of my own volition. Of course, at this relatively young age, everyone is at a different level of maturity. And people decide to accept the appointment to West Point for various reasons—the quality education, to please their parents, and so on. If they base the decision on a desire to serve the country, and they enjoy working with people, then I think that the experience will most likely be compatible with their goals. Because one is constantly learning more about the military profession during the four years there, the reasons for entering West Point are often much different from the motivation for staying. When asked why he came to West Point, a friend of mine replied, "Just to check it out." He served with distinction and was a great officer.
Tom Clancy:	How was the West Point curriculum overall?
H. R. McMaster:	The broad-based education was very beneficial. The curriculum includes a solid grounding in math, sciences, and engineering, as well as English, history, and the social sciences. It also includes military science courses that familiarize cadets with the Army organization, historical tradition, administrative procedures, and tactics. The wide range of courses in the core curriculum allows cadets to identify the discipline in which they have the greatest aptitude and interest. Because my interests and aptitude are not as great in math, science, and engineering as in other areas, these courses were a source of considerable consternation for me. My most formidable academic challenge came during the second [yearling] year. We took something like twenty-four credit hours, including Saturday classes. Electrical engineering, chemistry, and physics—all at the same time! I depended heavily on my classmates and learned how to study in a disciplined, organized, and efficient manner.
Tom Clancy:	How has the curriculum evolved since you graduated?

H. R. McMaster: When I went to West Point, cadets "concentrated" in a particular discipline. Mine was international relations, and I took complementary electives in history and Spanish. I was quite interested in Latin American studies and had the opportunity to visit Peru as an exchange cadet. The number of credits, however, fell short of that required to qualify as a "major." The current curriculum has preserved the broad-based curriculum, but also permits cadets to major in a field of their choice.

Tom Clancy: How was the quality of the instructors and curriculum?

H. R. McMaster: The instructors held graduate degrees from the nation's top universities, and were extremely dedicated and enthusiastic teachers. The largest classes had only twelve to fifteen students. And the instructors were willing to meet with cadets anytime during the day to assist them individually. The preponderance of the professors are Army officers, although some faculty members are civilian professors or officers from the other services [Navy, Air Force, etc.]. Because the instructors have an interest in each cadet as a future Army officer, they put an extraordinary amount of effort and energy into his or her intellectual and professional development.

Life at West Point is more than just academics and military discipline. On the contrary, it is also filled with comradeship and adventures. Following each school year, the cadets are given the opportunity to spend time with Army field units, or to study abroad. And no description of life at West Point would be complete without stories of athletics. H.R.'s experiences were no different, as he became a member of the Academy rugby team:

Tom Clancy: What else can you tell us about your experience at West Point?

H. R. McMaster: First of all, one makes friends for life. . . . As for enjoyment, I had the most fun playing on the Army rugby team. Our head coach, Major J.D.A. Baker [a British officer], was an outstanding example of what a military officer should be. We [the team] were a very close-knit group and the team provided a welcome respite from daily cadet life. We had a great time together and competed very successfully. And while playing rugby with the team in the fall of 1983, I met my wife, Katie [Trotter].

Tom Clancy:	Will you tell us how that happened?
H. R. McMaster:	In 1983, the Army-Navy game was held out in the Rose Bowl in Pasadena, California. The rugby team went out early to play some matches with West Coast clubs. I met Katie through a teammate who had become acquainted with her best friend. Katie is a graduate of the University of San Diego, and had been teaching high school. At the time, she was rather skeptical about meeting a cadet. Despite her predisposition, and an injury I had sustained in our game against California State University, Long Beach, we had an enjoyable time and established an immediate rapport. Afterward, we carried on a long-distance relationship and were able to see each other every month or two. We were married in July of 1985—about a year after I graduated from West Point.

During the cadets' four years at West Point, they are exposed to a variety of military experiences designed to assist them with making the vital decision of what branch of the Army (Armor, Infantry, Intelligence, etc.) they will join and specialize in during their career. Almost everyone who follows this path seems to have a story about how he wound up where he did, and H.R. is no exception:

Tom Clancy:	At what point did you decide what branch of the Army you wanted to go into?
H. R. McMaster:	Cadets decide on their branch during the senior ["Firstie"] year. When I entered West Point, there was no doubt in my mind that I would pursue a commission in the infantry. The summer between my sophomore ["yearling"] and junior ["cow"] year reinforced this desire: I went to Fort Carson, Colorado, to the 1st Battalion, 8th Infantry Regiment [usually abbreviated as the "1st of the 8th"]. For a five-week period I was assigned as a mechanized infantry platoon leader. During Cadet Troop Leader Training (CTLT), a cadet usually "shadows" a platoon leader as an intern. I was very fortunate that the 1st of the 8th was short a platoon leader, and the battalion commander gave me full responsibility for running the platoon. It was a great opportunity for me to experience the relationship between a new officer and the soldiers of a platoon, particularly in establishing a rapport and earning the respect of the men. I also went with the platoon to the newly opened National Training Center [NTC] at

Fort Irwin, California. And thanks to the platoon's non-commissioned officers [NCOs, i.e., sergeants] and the other officers in the company, I learned a great deal about small-unit leadership and operations.

During my senior year, Army Aviation became an independent branch. I became interested in flight school then, and decided to choose Aviation as my branch. I had read about air cavalry operations in Vietnam, and wanted to become an aero scout. After graduation I attended the Airborne School at Fort Benning, Georgia, and the Armor Officer Basic Course [AOBC] at Fort Knox, Kentucky, on the way to flight school. At AOBC, a flight physical revealed a previously undetected astigmatism in my eye that disqualified me from Flight School. As a result, I was immediately transferred to Armor Branch. To prepare myself better for leadership of a ground combat unit, I enrolled in pre-Ranger training and competed for an assignment to Ranger School. Before departing Fort Knox, I met with an assignment officer to discuss my request for an assignment in Germany. The officer, a captain, recommended that I go to Fort Hood, Texas, because the 2nd Armored Division was the only fully modernized division in the U.S. Army. The "Hell on Wheels" Division [the 2nd Armored's nickname] had recently received the M1 Abrams heavy tank and the M2/3 Bradley fighting vehicle. I took the officer's advice.

Tom Clancy: How long was the AOBC at Fort Knox, and what did you learn?

H. R. McMaster: I entered in July and finished in November. AOBC focuses primarily on ensuring that new lieutenants are able to employ a tank platoon [a unit of four tanks and sixteen men] effectively in various tactical situations. The Armor School ensures that tank officers are technically competent on the tank's systems, its weapons, and its communications equipment. Students also become familiar with Army administration, supply, and maintenance procedures. The capstone exercise is a week-long "war," during which the new lieutenants, organized into tank platoons, battle each other under the scrutiny of the School's instructors.

Tom Clancy: You mentioned that you had applied to Ranger School. Please tell us what that was like.

H. R. McMaster:	Ranger School is primarily a leadership school. The course puts students under intense physical and mental strain, and evaluates how they operate as leaders of teams while exhausted and under pressure. Students also learn tactics—specifically how to plan and lead dismounted patrolling operations. And they quickly discover that a well-conditioned human body can take a surprising amount of punishment. They are exposed to the elements for extended periods and operate in a variety of difficult environments—such as mountains, deserts, and jungles [swamps]. Ranger candidates receive only two meals a day and lose a great deal of bodyweight.
	Ranger School is invaluable. It reinforces the preeminent importance of teamwork—no "individual" students graduate. Your "buddies" help you when you need it, and you return the favor when you can. And if the teams of students don't operate well together, the Ranger instructors make it harder on everyone.
Tom Clancy:	What did you do after graduation from Ranger School [in March 1985]?
H. R. McMaster:	I drove from Fort Benning, Georgia, to Fort Hood, Texas, and reported for duty to the 1st Battalion of the 66th Armored Regiment, 2nd Armored Division, in the Headquarters Company. My first assignment was support platoon leader [supply and transportation]. After eight months I was transferred to B Company as an M1 tank platoon leader.

When it first arrived in the U.S. Army inventory in 1981, the M1 Abrams was a whole new world to the young crews who had to learn to use its complex systems. Let's look at H.R.'s perspective on this.

Tom Clancy:	Explain the difference between the M1 Abrams tank and the older M60A3 Patton.
H. R. McMaster:	The M1 was a tremendous leap forward in maneuverability, accuracy, and protection. It fires on the move with no problem, and the suspension system provides an exceptionally smooth ride. The fire-control system automatically computes the lead for engaging moving targets. The thermal-imaging system turns night into day. And the 1,500-horsepower engine is powerful and responsive.

The fire-on-the-move capability is a vast improvement over the M60.

Tom Clancy: Could you please try and tell us about the tank platoon that you commanded (in the 1st Battalion of the 66th Armored Regiment, 2nd Armored Division) during this time?

H. R. McMaster: A U.S. tank platoon consists of four tanks, and is manned by one officer [a second lieutenant] and fifteen non-commissioned officers and soldiers. Sergeants command the tanks, except for the platoon leader's tank. A platoon is capable of operating in two sections of two tanks each, one commanded by the lieutenant and the other by the platoon sergeant, usually a sergeant first class. The platoon leader focuses primarily on the tactical employment of the unit; and the platoon sergeant ensures that the platoon is well supplied and prepared for operations. These responsibilities, however, are not clearly demarcated. A good relationship between the platoon leader and the platoon sergeant, born of mutual respect and shared goals, is the key to building a cohesive, effective team. Generally thirty-five to forty years of age, the platoon sergeant is considered a technical expert on his vehicles, but he also must have the ability to train, care for, and motivate soldiers. If you have a sharp platoon sergeant, you can just about guarantee an excellent platoon.

Tom Clancy: Explain the crew positions inside the M1 Abrams tank itself. For example, because it's the easiest job, do you always give the loader's job to the least experienced person?

H. R. McMaster: No, the loader is *not* necessarily the easiest job. Each position on the tank is crucial. Loader is probably the best position from which to learn about how to maintain and operate the tank. The normal career progression is from loader, to driver, to gunner, and then to tank commander. So a loader has the opportunity to really be an apprentice. For example, he learns the automotive side of a tank from the driver. When the driver works on the hull or automotive and suspension part of the tank, the loader works with him. From the loader's position inside the turret, he gains expertise from the gunner and tank commander while he is doing his own job. He is really in the best position to gain familiarity with all aspects of the tank. The driver focuses primarily on ensuring that the tank is in proper running condition. The gunner maintains the 120mm main gun, 7.62mm coaxial machine gun, and the tank's fire-control system. The tank commander is responsible for the .50-caliber machine gun and supervises the efforts of his fellow crew members. I've oversimplified this description. None of the duties are set in stone, and the whole crew works together to get the job done.

Tom Clancy: Tell us about the thermal-imaging sights on an M1A1 tank like the one you used in Desert Storm.

H. R. McMaster: Just one thermal sight; the gunner controls it. He can switch between thermal [heat signature] and daylight [visible light] sights. But the tank commander has what is called a Gunner's Primary Site Extension [GPSE], through which he sees exactly what the gunner sees. And the tank commander can direct the gun and the sight from his position by using an override handle. The driver has a series of periscopes or vision blocks that he looks through. He can mount a night-vision device that amplifies ambient light onto a green field of view, so that he can also see at night.

By the end of 1987, H.R. had come a long way in the Army, and was beginning to think about how he might gain command of a small unit in Europe. After a year as a tank company executive officer, he was assigned as the battalion scout platoon leader. This was a significant move, for it headed him onto a path to his goal, command of a cavalry troop in Germany.

Tom Clancy: What happened to you going into the spring and summer of 1987?

H. R. McMaster: I was still a tank company executive officer, learning more about the M1 tank and deepening my experience in small-unit operations. About this time, though, I had decided that I really wanted to be a scout platoon leader. I heard that the position was coming open, and I asked my battalion commander to consider me for the job. I couldn't have asked for a more fulfilling and rewarding experience. A scout platoon leader is responsible for six M3 Bradley scout vehicles. Five soldiers are assigned to each Bradley; the driver, the gunner, the Bradley commander [who is also a squad leader], and in the back, two scouts/observers. The general mission of the scout platoon is to perform reconnaissance and security for the battalion. The platoon focuses on finding or detecting the enemy early, to give the battalion commander time and information to help gain the most advantageous position over the enemy.

Tom Clancy: How many soldiers and vehicles are in a scout platoon?

H. R. McMaster: Thirty troopers all together. The leadership consists of a platoon leader [a first lieutenant] and a platoon sergeant. The platoon is organized into three sections. The headquarters section typically has the Bradley platoon leader and platoon sergeant in it. The other two sections are led by staff sergeants, who each control a pair of M3s. The platoon can also operate as two sections of three Bradleys each, with the platoon leader and sergeant splitting up and each controlling a section. It's a very flexible organization. In my own platoon, and later in Eagle Troop, we operated with the platoon leader and platoon sergeant split up, regardless of what formation we were using, to spread the platoon leadership out across the area of operation.

Tom Clancy: Since the M3 Bradley Cavalry Fighting Vehicle is quite different from a tank, please describe it and its armament.

H. R. McMaster: The M3 Cavalry Fighting Vehicle (CFV) is powered by a turbocharged diesel engine. The main armament on the vehicle is a 25mm automatic cannon called a "chain gun" or "bushmaster" [after the snake]. It fires two types of ammunition. One type of round is an armor-piercing bullet. The other is a high-explosive incendiary tracer round that explodes on contact with the target. The Bradley is also equipped with a 7.62mm coaxial machine gun, meaning the gun is mounted and moves together with the 25mm chain gun. A two-round TOW missile launcher gives the Bradley an anti-tank capability. To fire TOW missiles, the Bradley must halt on fairly level ground. The gunner acquires targets through either a daylight sight or a thermal-imaging sight similar to the M1A1 tank's.

Tom Clancy: In addition to the Bradley's mounted weapons, what other weapons would you normally carry?

H. R. McMaster:	Everyone carries his personal M16A2 assault rifle, and the dismounted scouts crew an M60 machine gun. The Bradley also carries AT4s [the U.S. Army replacement for the venerable M72 LAWS rocket], which are light, shoulder-fired, anti-tank weapons. The AT4 is quite accurate, and gives dismounted scouts a close-in anti-armor capability.
Tom Clancy:	When you moved over to the scouts, did you feel that you had finally found out where you were supposed to be going in the Army?
H. R. McMaster:	I really did. I enjoyed it thoroughly, and I couldn't have asked for a better team of soldiers to work with. Serving with that scout platoon was a great experience, and the training particularly at the National Training Center gave me a solid grounding in small unit reconnaissance and security operations.

As part of his growing experience, H.R. was given, in 1987, the opportunity to take his scout platoon to Exercise REFORGER. This gave him an appreciation of relatively independent operations, and increased his enthusiasm for gaining a troop-level command in Europe.

Tom Clancy:	Going into late 1987, what was happening to you at that time?
H. R. McMaster:	In late 1987, I got the opportunity to go to Europe during Operation REFORGER with the 2nd Squadron, 1st Cavalry Regiment, which was the division cavalry for the 2nd Armored Division. We went as a fourth scout platoon in one of the ground troops. At that time, division cavalry squadrons [the ones assigned directly to armored and mechanized divisions] were normally organized with only three scout platoons. Colonel Tom Dials, who later became chief of cavalry tactics at Fort Knox, was the squadron commander. He was a great leader and tactician and I learned a great deal from him. His squadron was top-notch, and it was an honor to serve with them.
Tom Clancy:	What exactly is Operation REFORGER?
H. R. McMaster:	It stands for "Return of Forces to Germany." During the Cold War, there were forces designated to reinforce units

already stationed in Germany. REFORGER provided an opportunity to rehearse deployment and large-scale operations. During REFORGER, we conducted operations across wide frontages and in great depth—operations that one doesn't get the opportunity to practice in the United States. We also got to see a lot of North Germany.

During REFORGER, units trained in a variety of places with the forces of many nations. And the missions that they trained for were not all defensive. The new U.S. Army maneuver doctrine—first spelled out in the 1982 edition of the Army field manual FM 100-5—emphasized that even missions that are primarily defensive involve offensive operations. This was a shift in conceptualization for Army units. And units of all sizes quickly changed their focus to offensive operations. REFORGER-87 allowed H.R. and his scout platoon to participate in corps-sized offensive maneuvers:

Tom Clancy: What kinds of things did you do during REFORGER?

H. R. McMaster: Well, we received one mission from Colonel Dials, to conduct reconnaissance deep behind enemy lines in order to find a landing zone [LZ] for a corps air-assault [helicopter] operation. We found the area heavily defended and warned the infantry brigade that was scheduled to be inserted. Other operations included screening forward of a large armored formation, to destroy enemy security forces and locate the enemy's main defensive positions.

Tom Clancy: What did you learn from the operation of larger units during REFORGER?

H. R. McMaster: One of the most important lessons is that information one could easily overlook might be extremely valuable to the main body of the attack formation. It is vital for large armored units, such as brigades and divisions, to have detailed information about the zone through which they are moving, so they can keep the forward momentum during an offensive operation.

Tom Clancy: During REFORGER, how did you get this information to other units in the division?

H. R. McMaster: We reported the information to Squadron Headquarters, but also used what we call "connecting files"—i.e., when the unit behind a cavalry element [such as a tank or mechanized infantry battalion] pushes forward its own reconnaissance element [usually a scout platoon] to connect with reconnaissance units forward of them. It is the fastest and most efficient way to pass on the information that you have gathered. Cavalry units can also provide guides to meet following units on the ground, to give them more detailed information. Communication between the attacking force and the cavalry forward of them is vital, and the units monitor each others' radio frequencies.

Tom Clancy: What type of communications equipment did you have at your disposal?

H. R. McMaster: Each scout track [M3 Bradley] had two FM radios. Scout squad leaders had responsibility for monitoring an adjacent unit or a unit to the rear. The radios are outfitted with speech security equipment that uses an encryption code to scramble transmissions in such a way that only other radios that have that particular code can receive them. This prevents the enemy from eavesdropping. Our platoon also had three additional radios mounted in the back of the Bradley capable of sending text messages [no use of one's voice is required] by typing the message in on a keypad. It takes very little time to transmit [called a "burst" transmission], so the enemy cannot jam or direction-find the transmission. Each Bradley also had man-pack portable radios to communicate with dismounted scouts.

Tom Clancy: With REFORGER as your first taste of Europe, what were your expectations if you had been forced to fight in a major war?

H. R. McMaster: As a junior officer, you focus on ensuring that your unit is prepared for combat. I was very confident that we were ready. Preparing for battle was the focus of the entire Army. I knew that I was part of an organization committed to excellence. The Soviets had a large and powerful military, but we knew that we had a qualitative edge in people, organization, equipment, and training.

By the end of 1987, it was time for H.R. to begin moving up to the "middle management" of the U.S. Army. To this end, he attended the Advanced Armor School at Fort Knox, Kentucky, and prepared to move to his first European assignment with the 2nd Armored Cavalry Regiment in Germany:

Tom Clancy:	After REFORGER, what did you do next?

H. R. McMaster:	In early 1988, I attended the Armor Officer Advanced Course at Fort Knox, Kentucky. It was the first time in quite a while that I knew that I was going to have weekends off and have more time to spend with my family [H.R.'s and Katie's first daughter, Katharine, had been born in September of 1986]! As for the course itself, it was a seminar-type environment in which the students focused on planning large scale [regimental/brigade-sized] operations and worked their way down to company-level operations. In the curriculum, we took a look at possible scenarios in Korea and some portions of North Africa, because of the varied terrain and political volatility of the regions. In the Armor course, one learns a lot from his colleagues, by comparing experiences and ideas.

Tom Clancy:	What did you do after you completed the Advanced Armor course?

H. R. McMaster:	I had been spoiled by my experience as a scout platoon leader, and I was anxious to stay in the cavalry community [there were three regiments at the time, the 2nd, 3rd, and 11th]. Back in 1987, Colonel Dials recommended that I seek an assignment to the 2nd Armored Cavalry Regiment. And that is what I got; my assignment was to the regimental staff of the 2nd Armored Cavalry Regiment [2nd ACR], which was based in Bavaria [in southern Germany]. The Regimental Headquarters was in Nuremberg, with the 1st Squadron in Bindlach, which is about a forty-five-minute drive [northeast, toward the border] from the Headquarters. The 2nd Squadron was based in Bamberg, and the 3rd Squadron in Amberg, while the 4th Squadron [the air cavalry squadron] was closer to Nuremberg, in Feucht.

Tom Clancy:	What, at that time, was the mission of the 2nd ACR?

H. R. McMaster:	The 2nd ACR was the cavalry regiment of the VII Corps, which was prepared to assist in the defense of Central Europe. Each corps has a cavalry regiment, which acts as the "eyes and ears" of the corps commander. The 2nd ACR was prepared to execute what is called an "economy of force" mission. The regiment prepared to defend a wide sector, so the corps commander could concentrate the

preponderance of his force where he expected the enemy's main effort. The regiment was reinforced with an infantry battalion task force [with M2 Bradley Infantry Fighting Vehicles] and a tank battalion task force. The regiment was a busy outfit; and, while preparing for combat, also patrolled a portion of the West German/East German and West German/Czechoslovakian border. With the end of the Cold War in the late 1980s, the 2nd ACR de-emphasized border surveillance and control, and focused more exclusively on training hard to be ready for combat. We could still maneuver extensively in the German countryside, and accomplished some invaluable training.

In early 1990, H.R. finally had the opportunity to command his own maneuver unit, Eagle Troop of the 2nd ACR. He and Katie also had the excitement of the birth of their second daughter, Colleen, in February 1990. Nine months after he took over Eagle Troop, his unit was alerted for what would become a combat mission, Operation Desert Shield. Even before the alert orders came down, he had a feeling of impending action, and began to "work up" the personnel of Eagle Troop to get them ready for combat.

Tom Clancy: 1990 arrives, and what are you doing?

H. R. McMaster: I remained on the 2nd ACR staff until January, 1990, as the chief REFORGER planner for the regiment. In the plans office, which included myself, another captain, a staff sergeant, and two specialists, I drafted the REFORGER plans [the exercise is run yearly] and coordinated with the VII Corps staff. I learned a great deal from our Regimental Commander, Colonel L.D. Holder, a military historian and coauthor of *FM 100-5*. After REFORGER, I left that job and began the process of taking command of Eagle Troop in the 2nd Squadron at Bamberg. Also my second daughter, Colleen, was born in February 1990; and one month later we moved to Bamberg.

Tom Clancy: How was the lifestyle for you and your family in Bamberg?

H. R. McMaster: We enjoyed living in a small German community and made some close friends. Although I was gone much of the time, my family and I got to see a lot of Europe. My wife, Katie, taught BSEP, which is an adult education program for soldiers.

Tom Clancy: What were you and Eagle Troop doing when Iraq invaded Kuwait in August of 1990?

H. R. McMaster: We were in the field for gunnery and maneuver training. I was pressing the troops pretty hard to fine-tune our ability to move and shoot as a team. We had completed a very successful exercise at the Combat Arms Maneuver Training Center [the European equivalent of NTC] in May, but had some new personnel in key positions. As we came back from the gunnery exercise and started getting ready for a tactical operation, the news came in that Iraq had just invaded Kuwait. I spoke to the troop about the possible consequences of this event and told them to maximize the training opportunity because the next order they received might be in the desert of Saudi Arabia. As it happened, the first troops to go into Saudi Arabia [the 82nd Airborne and 101st Air Assault Divisions] came from the U.S.-based Rapid Deployment Force [RDF] instead of the European-based VII Corps.

Tom Clancy: What was the mission of the 2nd ACR going to be during this deployment?

H. R. McMaster: 2nd ACR was deployed as part of the forces [VII Corps] designed to give General Schwarzkopf [the Central Command commander] the offensive punch to eject the Iraqis out of Kuwait. Initially the regiment was to provide security for the rest of VII Corps, which was moving down from Germany. In addition to the 2nd ACR, VII Corps eventually included the 1st and 3rd Armored Divisions, the 1st Infantry Division (Mechanized), and the British 1st Armoured Division.

Tom Clancy: Prior to the alert order in November of 1990, had there been any contingency planning for the 2nd ACR to deploy to Saudi Arabia?

H. R. McMaster: There had been quite a bit of rumor and speculation. We discussed the likelihood of deployment and the general nature of offensive operations in the Iraqi desert, but did no deliberate planning at the squadron level until we were notified.

Tom Clancy: Could you describe what happened when the alert order came?

H. R. McMaster:	Our unit [2nd Squadron, of which Eagle Troop was a part] was the first unit of the 2nd ACR to go [to Saudi Arabia]. Though we had about twenty-four-hours' notice before we had to get all the equipment packed up and ready for shipment [to the embarkation ports], we were prepared, because our mission in Germany was to deploy locally with minimal notice. The vehicles were pretty much in top shape, and went out on a train for transshipment to the port of Bremerhaven.
Tom Clancy:	Once your vehicles were gone, did you concentrate your energies on helping the other squadrons in 2nd ACR prepare their equipment?
H. R. McMaster:	Not really. We actually concentrated our efforts on getting our own troops ready for the move. We also focused on training in a number of areas, from basic survival techniques and chemical defense to desert navigation and vehicle maintenance. In addition, we attended and gave briefings on the Iraqi Army and enemy situation in the Kuwaiti Theater of Operations. And we prepared a manual prior to deployment called "FM 100-Eagle" [a reference to the U.S. Army's FM 100-5 field manual]. It went to all Eagle Troop leaders, and focused on information such as desert survival, first aid, prevention of heat injuries, driving techniques, Iraqi Army tactics, and so on. I had the opportunity to talk to all of the troop's soldiers in small groups about the nature of armored combat in the desert, drawing heavily on the World War II experience in North Africa and the Arab-Israeli Wars. We also talked about how we intended to modify our tactics—drills and formations—for use in the desert.
Tom Clancy:	What kind of special threats were you expecting from the Iraqis if they had attacked you in Saudi Arabia?
H. R. McMaster:	Nerve gas. We knew that this is what they had the most of. Though we thought the likelihood of being hit with it was small, we were well prepared to defend against it.

As we move into the Persian Gulf with Captain McMaster, it is useful to get to know something about the people and equipment that he took to war. In the story that follows you can see in microcosm what all of the troops of Operations Desert Shield and Desert Storm went through as they counted down the time to G-Day (February 24th), the start of the ground war.

Tom Clancy: Were you able to take all of your personnel from Eagle Troop with you?

H. R. McMaster: Yes, no one was left behind.

Tom Clancy: Can you tell us about your first sergeant in Eagle Troop?

H. R. McMaster: I had an exceptionally talented and effective first sergeant in William Virrill. First Sergeant Virrill set high standards for himself and the unit, and was able to establish a close yet professional rapport with the young soldiers. The first sergeant and I were in general agreement about the unit's goals and priorities and had a close working relationship. We were, in effect, partners.

Tom Clancy: Could you describe the different platoons in Eagle Troop, what they were equipped with, and their leadership?

H. R. McMaster: We had two scout platoons, with six M3A2 Bradley fighting vehicles each, and two tank platoons, each equipped with four M1A1 Abrams "heavy" tanks with the new [depleted-uranium] armor. The troop executive officer was Lieutenant John Gifford [West Point, 1987], and he operated out of the troop command post [an M577]. The 1st Scout Platoon [six M3A2s] was commanded by Lieutenant Mike Petschek, a Georgetown University graduate. The 1st Scout Platoon sergeant was Staff Sergeant Robert Patterson. The 2nd Tank Platoon [four M1A1s] was commanded by Lieutenant Mike Hamilton, from Norwich University; and his platoon sergeant was Sergeant First Class Eddie Wallace. The 3rd Scout Platoon [six M3A2s] leader was Lieutenant Tim Gauthier, who left the 2nd Platoon to become 3rd Platoon leader. He was a graduate of Arizona State University. His platoon sergeant was Staff Sergeant David Caudill. The 4th Tank Platoon [four M1A1s] was lead by Lieutenant Jeff DeStefano, a West Point graduate. His platoon sergeant was Staff Sergeant Henry Foy. I had a great deal of respect for the leadership of the troop. The officers and NCOs were exceptionally talented and dedicated leaders who genuinely cared for the soldiers in their charge. We seemed to complement each other in temperament and style. I consider them to be among my closest friends.

Tom Clancy:	How was the Eagle Troop's Headquarters Platoon organized and equipped?
H. R. McMaster:	The Headquarters Platoon included the mortar section [two M106 4.2-inch-mortar carriers] and the troop's maintenance, communications, and supply functions. Personnel in the Headquarters Platoon included about twelve mechanics, eleven mortarmen, and approximately twenty other communication specialists, supply personnel, medics, and support soldiers. The vehicles in the maintenance section included an M88 recovery vehicle, an M113 armored personnel carrier, and a pair of M35A2 2-1/2-ton trucks. One of these was configured as a tool truck, and the other stored the repair parts typically carried with a cavalry troop. In addition to the eight tanks in the platoons, I had an M1A1 tank [call sign Eagle-66, nicknamed "Mad Max"]. The first sergeant had another M113 assigned to him. The troop commander and first sergeant also had HMMWVs, which we converted into ambulances. The troop command post had an M577 mobile command post [based on an M113 chassis] and a 5-ton truck for carrying supplies. We had an additional HMMWV assigned to the troop executive officer. So, when we deployed, we had the following equipment in the troop: nine [M1A1 Abrams] tanks, thirteen Bradleys, two mortar tracks [M113 APCs with 106mm mortars], one M88 recovery vehicle, two organic M113 APCs configured as ambulances, an M577 mobile command post, four HMMWVs, a 5-ton truck, and two 2-1/2-ton trucks. Later, we received a FIST-V fire-support vehicle [an M113 APC configured to call for

Captain H. R. McMaster (center) with his Eagle Troop platoon leaders. (Left to right) 1st Lt. Jeffery DeStefano, 1st Lt. Timmothy Gauthier, 2nd Lt. Michael Hamilton, and 1st Lt. Michael Petschek.

H. R. McMaster

and adjust artillery fire], with an excellent fire-support team led by Lieutenant Daniel Davis. Our radio equipment was pretty standard. We did have one of the new NAVSTAR Positioning System [GPS] units that we used in the lead scout platoon on Mike Petschek's track.

Tom Clancy: Were there any last-minute changes before departing for Saudi Arabia?

H. R. McMaster: Not really. When we arrived, the timing was perfect. Our vehicles and equipment got there the day after we did.

Tom Clancy: What was it like saying good-bye to your family?

H. R. McMaster: Of course I would miss Katie and the girls tremendously. But we did not have a particularly dramatic emotional moment. Both Katie and I believe that it is best to keep a positive outlook. That meant that I did not want my children—my oldest daughter, Katharine, was only four and a half years old—to know we were going into combat. So Katie and I told them I was going to the field as usual. And we didn't take them to the going-away ceremonies. We thought that the sight of other families saying good-bye—with the waving of flags, and crying—might disturb our young daughters.

Tom Clancy: What was it like to see others departing from their families?

H. R. McMaster: I think it was more difficult for the families than for the soldiers. The troops knew there was a mission to accomplish and were leaving as members of a close-knit team. The wives and children had to cope with the uncertainty surrounding the deployment. Yet the wives supported one another, and I think that the soldiers' confidence was reassuring to them.

Tom Clancy: So then you went directly to Saudi Arabia?

H. R. McMaster: Right. We took commercial flights out of snow-covered Nuremberg, flew into Dhahran, and bused in to the port facility at Al Jubail.

When he reached Saudi Arabia in early December 1990, H.R.'s first job was to get Eagle Troop's equipment off the ships at Al Jubail, and get them to their first assembly area along the Saudi/Iraq border. After that, his early days were spent keeping his troops healthy, fit, and fed, keeping the troop's equipment ready for action, and getting his personnel trained and positioned for the coming assault into Iraq.

Tom Clancy: How was the weather when you got there?

H. R. McMaster: When we landed on December 4, 1990, it was very hot, in the mid-90s [Fahrenheit]. Although it wasn't nearly as hot as it had been in the late summer, it was a drastic change from cold, snowy Germany.

Tom Clancy: At the port of Al Jubail, what was necessary on your part to get your vehicles off-loaded and ready to move?

H. R. McMaster: We drove them off the ships, marshaled them in the port area, and then road-marched to the staging area. After that, we spent several days getting our equipment together and painting the combat vehicles sand tan. We then loaded the vehicles onto commercial heavy equipment transports [HETs—"low-boy" tractor-trailer rigs] and moved to our tactical staging area, where we would remain until after the air campaign began.

Tom Clancy: Where was your first setup area and what other units were around you?

H. R. McMaster: East of the Wadi al-Batin and north of the Tapline Road. Generally, it was northeast of the huge military complex at King Kalid Military City [called KKMC for short]. At first, no other units were near us—we were west of the Marines and the 3rd ACR. Our first assembly area was in an absolutely flat, featureless, and uninhabited portion of the desert.

Tom Clancy: When did the entire regiment get into Saudi Arabia, and what did you do at that time?

H. R. McMaster: The rest of the regiment had arrived by mid-December. The 2nd Squadron S-3 [operations officer], Major [now Lieutenant Colonel] Douglas MacGregor, developed a training plan during which we focused sequentially on individual, crew, platoon, and troop-level tasks. We then maneuvered the entire squadron of three troops, a tank company, a headquarters company, and a howitzer battery as a single unit. The squadron exercises were challenging and emphasized night operations. We also rehearsed our march formations and battle drills, to prepare for leading the VII Corps into Iraq. Because 2nd Squadron was the first complete unit of the 2nd ACR to get into Saudi Arabia, we were able to gain a lot of desert-maneuver experience in a short time.

One of the most important late additions to the equipment of Eagle Troop, and all of the Coalition forces in the Persian Gulf, was a number of the new NAVSTAR GPS terminals, which greatly aided in desert navigation. Though the U.S. Army had about a thousand of these units prior to the Iraqi invasion of Kuwait, this number grew to several thousand as an emergency procurement of the car-stereo-sized units were bought and sent to field units. In addition, thousands of commercial GPS units were bought by individuals for use in everything from tractor-trailer trucks to helicopters:

Tom Clancy: When did you get the additional GPS terminals that you used during the ground war, and what were they like?

H. R. McMaster: In late December 1990, the regiment received a number of additional Trimble TRIMPACK GPS terminals, which could be powered off a battery or the power supply of a vehicle. They looked like portable car stereo units, and we mounted them on the top of the tank with velcro and foam rubber. They can be programmed to give position readouts using the military grid reference system [divided into one-kilometer/.61-mile squares], and are accurate to within just a few meters of ground truth. You can also program them with a series of "waypoints," to guide you as you move across the desert. We had four of these units in the troop, with one assigned to each of the command M3s of the 1st and 3rd Scout Platoons, one with my tank, and the other with Dan Davis' fire-support vehicle [FIST-V]. GPS gave us tremendous advantages, and we couldn't have operated as well as we did without them. Unfortunately there were only six satellites in the constellation at the time [a total of twenty-four are eventually planned], so we didn't have around-the-clock satellite coverage. We called the periods in which we couldn't receive signals [and thus lost the use of the GPS system] GPS "sad times." Whenever this occurred, we had to revert back to other systems, such as LORAN or dead reckoning. Because of the lack of terrain features in the desert, dead reckoning entailed dismounting crew members with a compass to line the vehicle up on a magnetic azimuth. Once the vehicle was in line, the gunner put the weapons system into stabilized mode, and the driver kept in line with the gun tube. Drivers measured distance using the vehicle odometer.

Tom Clancy: How long were the GPS sad times?

H. R. McMaster: About a maximum of 40 minutes. It happened to us once when we were repositioning between two other units in a sandstorm. Needless to say, it was a bad time for this to happen. But 1st Platoon scouts did a great job getting us to where we needed to go.

Along the way, H.R. had to deal with all of the things that U.S. military commanders have dealt with since George Washington at Valley Forge. Keeping people fed, Christmas away from home, keeping morale up. There was even an occasional encounter with friendly Bedouin tribesmen.

Tom Clancy: What was an average day like for Eagle Troop?

H. R. McMaster: First we established circular perimeters for 360° security. We conducted physical training [PT] at crew-level every morning—push-ups, sit-ups, and sprints. It's vitally important that soldiers remain fit, limber, and energetic. It reinforced cohesiveness at crew level. Breakfast was normally a T-ration type. For those who do not know, an A-ration is fresh food, procured locally, and supplemented with B-rations, which are in cans. A T-ration is a cafeteria-style meal, packaged in rectangular trays and heated in boiling water. We also ate commercially supplied Top Shelf and Beefaroni-type meals [from emergency Army procurements]. Because of their limited supply, we were saving the Meals Ready to Eat [MRE] field rations for the ground war. As far as personal hygiene, we had crudely fashioned but functional latrines and gravity-fed showers. At the end of the day, latrine refuse would be mixed with diesel fuel and burned. When we moved, these facilities did not go with us. And when we moved to the VII Corps staging area, we left our tents behind and lived on our armored vehicles. The mail kept up with us, though; and we received lots of much-appreciated support from people at home. Schools and businesses were particularly good about sending "any soldier" packages and letters. The mail really boosted morale.

The crew of Eagle-66 (left to right) Captain H. R. McMaster (commander), Staff Sergeant Craig Koch (gunner), Specialist Christopher Hedenskog (driver), Private 1st Class Jeffrey Taylor (loader).

H. R. MCMASTER

Tom Clancy: How was morale in general?

H. R. McMaster: Our spirits were very high the entire time. We kept busy training and were extremely confident. The troop really bonded as a unit and took on an almost familial character.

Tom Clancy: What was Christmas 1990 like for your men?

H. R. McMaster: Eagle Troop was given the mission to guard Logistics Base Alpha. When we came into the base in battle formation, many of the soldiers there had never even seen a tank before and greeted us enthusiastically.

On Christmas Day, the platoons rotated off duty for the Christmas meal, so each had a special dinner with turkey and all the trimmings. The troop wives back in Germany had individually wrapped two presents per soldier, including personal hygiene items, stationery, and candy. We even had a little Christmas tree that lit up. The first sergeant and I put the tree on the front of the HMMWV and drove the troop perimeter to deliver the presents. We also spent some time with each of the soldiers. And our chaplain held several separate services so every soldier could attend.

Tom Clancy: How did you react to—or interact with—the local Bedouin people?

H. R. McMaster: We were briefed on their local customs and traditions and respected them. But our contacts with the indigenous population were very limited until later, when we helped secure southern Iraq after the cease-fire.

Tom Clancy: Did you listen to the radio a lot?

H. R. McMaster: Yes, we listened to the Baghdad station, which was the soldiers' favorite—with "Baghdad Betty." The Iraqi propaganda from there was unsophisticated, ridiculous, and pretty funny. They did, however, play decent music. On the downside, it seemed that every time we moved, we would be just outside the reach of Armed Forces Radio. BBC World Service radio was a good source of timely news. We didn't have CNN, though!

On January 16th, 1991, Operation Desert Storm kicked off with a massive aerial bombardment of Iraqi targets and forces. As they watched the explosive storm to their north, and suffered an early scare, they continued to prepare for the coming assault upon the Iraqi forces in Iraq and Kuwait.

Tom Clancy: When the air war started, what signs of war did you see?

H. R. McMaster: There were a lot of aircraft overhead; and through our night-vision goggles, we could see the northern horizon glowing. To celebrate the event, I awakened the cooks early that morning and asked them to serve pancakes and eggs cooked to order for breakfast. After that, I talked to the troop and sketched out the concept of the air campaign and reinforced some of the main aspects of the plan for the ground offensive. On the same day, we reacted to a false report of an Iraqi attack across the border. We were into our vehicles in seconds, cranked up and headed north to our covering-force area. Later, we were actually thankful that the false alarm happened, because it brought home to all the troops that this was no longer a deployment exercise, but an actual war.

Shortly after the air war started, 2nd ACR, along with the rest of the VII Corps and XVIII Airborne Corps, began a long movement several hundred miles to the west to support the "Hail Mary play" that was to be the centerpiece of General Norman Schwarzkopf's plan for the ground war phase (called Desert Saber) of Operation Desert Storm. Done in almost complete secrecy from the Iraqis (who had their intelligence collection limited to pirating signals off the CNN satellite feeds by this time), it was designed to allow the Coalition forces to cut off any possible escape routes from Kuwait into Iraq, and in particular, to allow VII Corps, with its heavy armored divisions, to destroy the five Republican Guard divisions that were standing by on the old Iraqi/Kuwaiti border to attack the Arab and Marine Corps units that were to liberate Kuwait itself. These divisions, oversized and equipped with the best equipment in the

Iraqi Army, were felt to be the primary threat to the forces attempting to liberate Kuwait. Assigned to VII Corps as their organic reconnaissance element, 2nd ACR was going to lead the way to the Republican Guard; then the rest of the corps was going to destroy it.

Tom Clancy: When did the move west to support the "Hail Mary" start?

H. R. McMaster: A few days after the air war started in late January 1991. After about two weeks, we moved north to secure the area just south of the Iraqi-Saudi Arabian border. And then during the week prior to the start of the ground war, we heard many rumors of late-breaking peaceful resolutions. It was important at this stage, however, that the soldiers remain focused on combat. I deliberately assured them that we would go into battle. Although we would have welcomed peace, it would, in some ways, have been a disappointment not to attack. We were ready, confident, and anxious to do our job.

Tom Clancy: What was the layout of VII Corps when you lined up to jump off on G-Day (February 24th, 1991)?

H. R. McMaster: The regiment moved forward of the corps, conducting what we call an offensive covering-force operation. The 1st and 3rd Armored Divisions were just behind us to our left [west] and right [east] respectively. In addition, the British 1st Armoured Division and the U.S. 1st Infantry Division (Mechanized) were getting ready to make their own push into Iraq and Kuwait to the east of us. The corps was to destroy the Republican Guards divisions that were occupying or defending Kuwait. The regiment was to move in advance of the corps, destroy enemy recon and security forces, locate the Republican Guard main defenses, and facilitate the passage of the heavy divisions through us to complete the destruction of the enemy. We expected to first encounter the Republican Guard just north and east of what we called Phase Line Smash.

The 2nd ACR attacked initially with two squadrons forward [2nd and 3rd], and one [1st Squadron] in reserve. The air cavalry [4th Squadron] screened forward of the ground squadrons. The regiment had been reinforced with a number of supporting units to increase its combat power. These included a battalion of AH-64A Apache and OH-58D Kiowa helicopters from the 1st Armored Division, the 82nd Engineer Battalion, two

155mm howitzer battalions, an MLRS [Multiple Launch Rocket System] battery, and a company of military police to assist with the processing of Iraqi prisoners. We attacked into Iraq on G-Day minus-one, the afternoon of 23 February 1991.

When Colonel Leonard D. "Don" Holder, the 2nd ACR's commander, led his regiment over the border berms (long mounds of earth) into Iraq, he commanded a unit greatly enlarged from its normal peacetime organization. In fact, by G-Day, he commanded a unit that was more like a small armored division, with all the cross-attachments assigned. The more specialized of these units, such as engineering and combat-intelligence units, would play a vital role in the initial assault against the Iraqis:

Tom Clancy: How did you prepare for the movement into Iraq on G-Day?

H. R. McMaster: Prior to moving into Iraq, the troop was reinforced with an engineer platoon that included an Armored Combat Earth Mover [called an ACE]. The ACE would cut a lane through the dirt berms that had been built by the Iraqis to demarcate the frontier with Saudi Arabia. The berms were not a formidable obstacle, but could impede the movement of large armored formations. One of the regiment's tasks was to make numerous cuts to ease the movement of the divisions behind us. I had the engineers construct several simulated berms to allow us to practice the breaching operation. When we executed the plan, the 1st Scout Platoon rushed forward with the engineers to secure and breach the obstacle. Once the breach was effected, my tank led the other tanks through in column. On the far side, the tanks deployed into a wedge formation. The 3rd Scout Platoon followed the tanks and established flank security to the west. Ultimately, 1st Platoon crossed and raced to resume the lead of the formation.

Tom Clancy: How good was your intelligence about what was on the other side of the berms in Iraq?

H. R. McMaster: Although an artillery prep was fired, we didn't expect any contact. The 3rd ACR [assigned to XVIII Airborne Corps to the west of VII Corps] had punched holes in the berm the day before; and Major MacGregor [the 2nd Squadron S-3] and I drove over to their area with our HMMWVs. After coordinating with them, we drove through their berm breaches and had a look at our squadron's crossing areas. We did not see evidence of any enemy activity. A week earlier, a 3rd Platoon dismounted patrol had checked the crossing areas at night and swept for mines.

On the night of February 23rd/24th, 1991, H.R. led Eagle Troop through the berms and into Iraq. Over the next couple of days, while the rest of VII Corps came through the berms and sorted themselves out, the 2nd ACR moved forward slowly. The weather was less than pleasant.

Tom Clancy: How did the actual breaching operation itself go?

H. R. McMaster: Lieutenant Ed Ketchum's engineers did a great job. They reduced the obstacle in less than a minute, and my tank went through seconds later. We wanted to be the first troop across, and made it with time to spare. It was actually a relief to be in Iraq. We were finally getting to do the job we had come to do. We test-fired our weapons on the move; and after a few minutes, the Cobras and OH-58s from the 4th Squadron took the lead, and Captain Tom Sprowls' Fox Troop took the point on the ground.

Tom Clancy: The weather conditions during the Ground War were bad, weren't they? Why don't you say a little bit about that?

H. R. McMaster: The first night, it poured rain. I slept in the tank commander's seat [in his M1, named "Mad Max"], hunched over with my helmet on the sight extension. On the 24th, though, the weather was generally good. But the night before our major fight, the 25th, the driving rains returned. The morning of the 26th, we encountered heavy ground fog, and later in the day the fog gave way to high wind and blowing sand that grounded aircraft and limited visibility.

One of the keys to the success of the ground phase of Desert Storm was the synchronization of every unit on the battlefield. This was necessary to help plan and execute air and logistics support operations, direct Coalition units to their desired targets (the Iraqi ground units), and to help avoid incidents of "friendly fire" or "fratricide."

Tom Clancy: Can you explain the movement scheme, particularly the concept of phase lines?

H. R. McMaster: Phase lines are graphic references laid out every few kilometers that are designed to keep track of forward progress during an offensive operation. Units report movement across a phase line to higher headquarters.

Tom Clancy: Tell us about the first two days of moving into Iraq.

H. R. McMaster: During the first two days of the campaign we covered short distances very rapidly, then halted for extended periods of time. Enemy resistance was very light and ineffective. Fox Troop made first contact with the enemy on the 24th. After a brief firefight, they captured large numbers of the enemy. Later that night we [Eagle Troop] detected a series of trench lines comprising an enemy infantry position. The Bradleys from 1st Platoon and my tank hit the position hard with 25mm fire, and TOW and tank HEAT rounds fired into the bunkers. The scouts also adjusted fire from the mortar section onto the area. The attack by fire drove the survivors toward Fox and Ghost Troops, who, after similar actions, took hundreds of the enemy prisoner. The next morning, as we moved further north, we encountered large numbers of surrendering enemy. We didn't have time to stop. So we ensured that they were unarmed and left them for units to the rear. . . . Many soldiers threw them food and water from the Bradleys, though.

Joe Sartiano's Ghost Troop had the only notable action in the squadron on the 25th. Ghost destroyed several MTLBs, captured several more, and drove them to the squadron command post. When we arrived for a meeting, the other troop commanders and I got to take them for a test drive. We knew that we had hit the security zone of the Republican Guard. The soldiers wore the Republican Guard insignia, and the unit had been well supplied and had new weapons and equipment. I told the troop not to assume that all the Iraqi units were as weak as those we encountered initially. I didn't want us to let our guard down.

February 26th, 1991, was the day for which the U.S. Army and H.R. had prepared themselves for over a decade: head-to-head armored combat with the best that the Iraqis had to offer, the Tawakalna Division of the Republican Guard. Of all the Republican Guards divisions that were encountered during Desert Storm, this was the only one which maneuvered with any real aggressiveness.

This was the way things were supposed to go: With Eagle Troop on the point, 2nd ACR was to locate the enemy, hold them at arm's length, and then

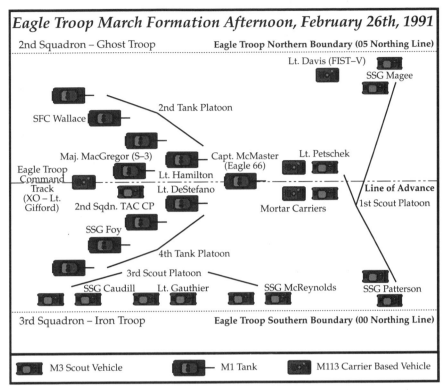

Eagle Troop March Formation Afternoon, February 26th, 1991

2nd Squadron – Ghost Troop

Eagle Troop Northern Boundary (05 Northing Line)

Lt. Davis (FIST–V)

SSG Magee

2nd Tank Platoon

SFC Wallace

Maj. MacGregor (S–3)

Capt. McMaster (Eagle 66)

Lt. Petschek

Eagle Troop Command Track (XO – Lt. Gifford)

Lt. Hamilton

Lt. DeStefano

2nd Sqdn. TAC CP

Mortar Carriers

Line of Advance

1st Scout Platoon

SSG Foy

4th Tank Platoon

3rd Scout Platoon

SSG Caudill   Lt. Gauthier

SSG McReynolds

SSG Patterson

3rd Squadron – Iron Troop

Eagle Troop Southern Boundary (00 Northing Line)

| M3 Scout Vehicle | M1 Tank | M113 Carrier Based Vehicle |

The formation of Eagle Troop, 2nd Cavalry Squadron, 2nd Armored Cavalry Regiment, on the afternoon of February 26th, 1991. The 1st Scout Platoon is out in front with Captain H. R. McMaster (Eagle-66) leading the wedge of M1A1 tanks. *Jack Ryan Enterprises, Ltd., by Laura Alpher*

pass the heavy armor of the 1st and 3rd Armored Divisions through to destroy the Iraqis. That was the plan. But the reality simply did not work out that way. High winds and blowing sand kept the helicopters on the ground, and it was left to the cavalry troopers on the ground to find the enemy, much like their mounted forefathers of the previous century.

Tom Clancy: When did you first encounter the first elements of Tawakalna Division of the Republican Guard?

H. R. McMaster: Eagle Troop first encountered reconnaissance elements of the Republican Guard on the morning of February 26th. Ghost Troop had destroyed two Iraqi armored personnel carriers, and a third BMP was taking evasive action. Staff Sergeant Patterson directed my tank onto the BMP, and my gunner, Sergeant Craig Koch, destroyed it with a HEAT round at long range—2,620 meters. As a result of this small action, I believe that the Iraqi scouts did not have the opportunity to warn their headquarters of our presence in the area. We remained in generally the same area throughout the morning of the 26th, and were disappointed when we received word that we would hold in position and assist the forward passage of the divisions that were closing to our rear. We were confident in our ability to fight and did not want to miss the opportunity to meet the enemy on the battlefield.

On the late afternoon of the 26th, as the regiment continued to move to contact with the Tawakalna Division, Eagle Troop encountered a small village astride the demarcation line, with 3rd Squadron to the south (below the 00 Northing or centerline of the VII Corps advance). After taking fire from machine guns and a dug-in ZU-23 anti-aircraft mount (a twin 23mm gun), the tanks and Bradleys of Eagle Troop responded in a much bigger way and silenced the enemy fire.

Tom Clancy: At what point did you come into contact with the village that gave you a problem?

H. R. McMaster: The regiment was controlling the operation very tightly. Our squadron, due to the narrowness of our zone, moved in a box formation, with Ghost and Eagle Troops forward and Fox Troop and the Tank Company in reserve. Eagle

The opening moves of the Battle of 73 Easting, Captain McMaster and the tanks of the 2nd and 4th Tank Platoons return the fire from the Iraqi village after taking several prisoners.

Jack Ryan Enterprises, Ltd., by Laura Alpher

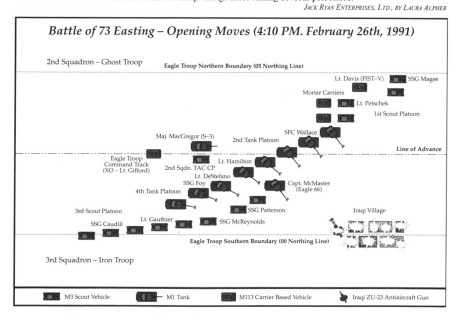

Battle of 73 Easting – Opening Moves (4:10 PM. February 26th, 1991)

2nd Squadron – Ghost Troop

Eagle Troop Northern Boundary (05 Northing Line)

Lt. Davis (FIST–V) SSG Magee
Mortar Carriers
Lt. Petschek
1st Scout Platoon
Maj. MacGregor (S–3) 2nd Tank Platoon SFC Wallace
Line of Advance
Eagle Troop Command Track (XO – Lt. Gifford) 2nd Sqdn. TAC CP Lt. Hamilton
Lt. DeStefano
SSG Foy Capt. McMaster (Eagle 66)
4th Tank Platoon
3rd Scout Platoon SSG Patterson Iraqi Village
SSG Caudill Lt. Gauthier SSG McReynolds
Eagle Troop Southern Boundary (00 Northing Line)

3rd Squadron – Iron Troop

M3 Scout Vehicle   M1 Tank   M113 Carrier Based Vehicle   Iraqi ZU-23 Antiaircraft Gun

Troop was positioned south of Ghost Troop and connected with Iron Troop [commanded by H.R.'s "Beast Barracks" roommate Dan Miller] of 3rd Squadron. Visibility remained limited, as morning fog had given way to high winds and blowing sand. The poor conditions grounded aircraft. At approximately 4:00 PM the regiment gave permission to move to the 67 Easting line. Staff Sergeant Reynolds' scout section encountered an Iraqi dismounted outpost; and as the scouts disarmed four Iraqi soldiers, the section came under enemy machine-gun and 23mm fire from a village about 1,200 meters further east. The scouts returned fire, and I ordered the troop's tanks to fire HEAT rounds into the buildings from which the Iraqis were firing. The tank fire blew large holes in the walls of the cinderblock buildings, collapsed roofs, and started fires. Almost immediately after suppressing the village, we received permission to advance three more kilometers to the 70 Easting line.

It was at this moment that all hell began to break loose ahead of Eagle Troop. Reports of enemy tanks (from Lieutenant Michael Petschek of the 1st Scout Platoon) ahead in the blowing dust and sand forced H.R. to make a decision: Did he stay where he was (as the accepted U.S. Army armored cavalry doctrine of the day suggested) and try to hold them there while the armor from the heavy divisions came up? Or should he assault the enemy armor?

When H.R. observed that after shooting three tanks in the early moments of the fight (in under ten seconds!) the enemy tanks were not returning effective fire, he made his decision. Realizing he had surprised and gained an advantage over the Iraqis, he ordered the troop forward to assault the enemy positions. Maneuvering and fighting with Eagle Troop at this time was Major MacGregor, the 2nd Squadron S-3 (operations) officer in an M1A1, and the 2nd Squadron Tactical Command Post (TAC CP) in an M2 Bradley:

Tom Clancy: What did you do then?

H. R. McMaster: I called up Red-1[1] [Lieutenant Petschek]—the 1st Scout Platoon leader's call sign —and told him to continue the attack to the 70 Easting line. He hesitated because, unknown to me, Staff Sergeant Cowart Magee's M3 crew had spotted an enemy tank [a T-72] and were about to engage it with a TOW missile. I ordered the troop's tanks to "follow my move," and took the lead of the formation in this uncertain situation. As Staff Sergeant Magee's gunner, Sergeant Moody, engaged the T-72, my tank crested an almost imperceptible rise in the terrain. I

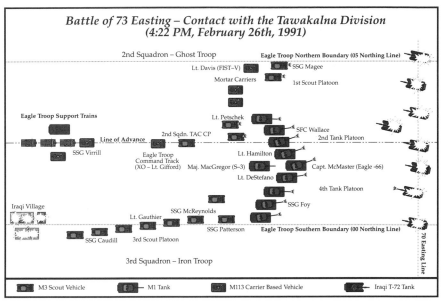

Battle of 73 Easting – Contact with the Tawakalna Division
(4:22 PM, February 26th, 1991)

2nd Squadron – Ghost Troop

Eagle Troop Northern Boundary (05 Northing Line)

Lt. Davis (FIST–V)

SSG Magee

Mortar Carriers

1st Scout Platoon

Eagle Troop Support Trains

Lt. Petschek

Line of Advance

2nd Sqdn. TAC CP

SFC Wallace

2nd Tank Platoon

SSG Virrill

Eagle Troop
Command Track
(XO – Lt. Gifford)   Maj. MacGregor (S–3)

Lt. Hamilton

Capt. McMaster (Eagle -66)

Lt. DeStefano

4th Tank Platoon

Iraqi Village

SSG McReynolds

SSG Foy

Lt. Gauthier

SSG Patterson   Eagle Troop Southern Boundary (00 Northing Line)

SSG Caudill   3rd Scout Platoon

70 Easting Line

3rd Squadron – Iron Troop

| M3 Scout Vehicle | M1 Tank | M113 Carrier Based Vehicle | Iraqi T-72 Tank |

Moving past the Iraqi village, the lead elements of Eagle Troop encounter a line of eight Iraqi T-72 tanks from the Tawakalna Division of the Iraqi Republican Guard. Captain McMaster in Eagle-66 and the Bradley fighting vehicles of the 1st Scout Platoon engage the enemy tank line, with the tanks of the 2nd and 4th Tank Platoons moving up to finish off the rest. Eagle Troop then moves forward to assault an Iraqi position behind the line of destroyed Iraqi T-72s.   *JACK RYAN ENTERPRISES, LTD., BY LAURA ALPHER*

immediately spotted eight enemy tanks in a defensive position oriented west (in our direction). The sandstorm had died down, and visibility had improved enough to see them with the naked eye.

Tom Clancy: What action did you take at this point?

H. R. McMaster: I reported "contact" to the troop and exhorted the two tank platoons [Green and White] to join me forward of the enemy position. At the same time, I gave my crew the command to fire, and my gunner, Staff Sergeant Craig Koch, destroyed three of the enemy tanks in under ten seconds. As we fired our third round, the tank platoons and Major MacGregor's tank joined in the fight. The fire distribution was perfect, and our nine tanks took out a large chunk of the enemy in the initial volley. I thought at the time that they were T-55 [an older model] tanks, because we were destroying them so quickly. The enemy return fire was also ineffective. Several T-72 tank rounds fell short, and their machine-gun fire had no effect on the armored vehicles.

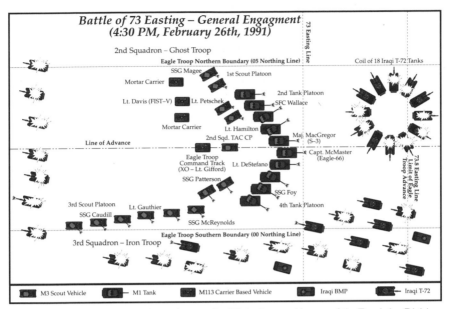

Battle of 73 Easting – General Engagment
(4:30 PM, February 26th, 1991)

2nd Squadron – Ghost Troop

Eagle Troop Northern Boundary (05 Northing Line)

73 Easting Line

Coil of 18 Iraqi T-72 Tanks

SSG Magee

1st Scout Platoon

Mortar Carrier

2nd Tank Platoon

Lt. Davis (FIST–V)   Lt. Petschek

SFC Wallace

Mortar Carrier   Lt. Hamilton

2nd Sqd. TAC CP

Maj. MacGregor (S-3)

Line of Advance

Capt. McMaster (Eagle-66)

Eagle Troop Command Track (XO – Lt. Gifford)   Lt. DeStefano

73.8 Easting Line Limit of Eagle Troop Advance

SSG Patterson

SSG Foy

3rd Scout Platoon   Lt. Gauthier

SSG Caudill

4th Tank Platoon

SSG McReynolds

Eagle Troop Southern Boundary (00 Northing Line)

3rd Squadron – Iron Troop

| M3 Scout Vehicle | M1 Tank | M113 Carrier Based Vehicle | Iraqi BMP | Iraqi T-72 |

Eagle Troop moves into a general assault on an Iraqi Brigade assembly area of the Tawakalna Division. The tank wedge of Captain McMaster, Major MacGregor (the 2nd Squadron S-3), as well as the tanks of the 2nd and 4th Tank Platoons, moves east to engage, followed by the rest of the troop. The advance terminates at the 73.8 Easting line.   *JACK RYAN ENTERPRISES, LTD., BY LAURA ALPHER*

As quickly as the line of tanks was destroyed, Captain McMaster sighted additional Iraqi armored vehicles beyond them and behind the village. This concentration of Iraqi armor was a defensive sector for a brigade of the Tawakalna Division. H.R. quickly ordered Eagle Troop to advance and engage the numerically superior enemy force which they had surprised.

Tom Clancy: What happened next?

H. R. McMaster: The Bradleys of the 1st Platoon tucked in behind the tank wedge to cover the rear, and 3rd Platoon was arrayed in depth to protect our open southern flank. The tanks focused on destroying enemy vehicles, and the Bradleys fired primarily at enemy dismounted infantry. As we penetrated the first line of enemy defenses, we began to engage additional enemy vehicles in depth further to the east and south. The enemy's main defense was centered on the 70 Easting line—our limit of advance. When John Gifford [the executive officer] reminded me that 70 was our limit of advance, I told him that we could not stop in the middle of the enemy position and would halt on the far side. In Germany, our regimental commander, Colonel [now Major General] L. D. Holder, had made it clear to us that junior officers were free to make decisions and take initiative based on the situation. We crested another ridge at approximately the 73 Easting line and entered the assembly area for the enemy's tank reserve. We destroyed the majority of the eighteen T-72s at close range, as they were trying to deploy against us. We halted just east of the reserve position.

Tom Clancy: As your attack slowed down, the battle was developing on your left and right, as Ghost and Iron Troops moved up and engaged the flank elements of the force you had been attacking. What happened?

H. R. McMaster: Ghost Troop [Captain Joseph Sartiano] engaged and destroyed the enemy forces to their front just as we halted at the 74 Easting line. Later, Captain Dan Miller's Iron Troop fought their way forward to narrow the gap between the two squadrons. We held in place, and later that night assisted the forward passage of the 1st Infantry Division (Mechanized).

Tom Clancy: How many vehicles did Eagle Troop destroy?

H. R. McMaster: Approximately thirty tanks, sixteen BMP infantry fighting vehicles, and thirty-nine trucks. The outcome of the battle I am most grateful for, however, is that Eagle Troop suffered no casualties. I thank God for that.

Tom Clancy: How long did the fight last?

H. R. McMaster: Approximately twenty-three minutes from when we were taken under fire from the village until we halted at the 74 Easting line.

A burned-out Iraqi T-72 on the 73 Easting battle-field. *Lieutenant Colonel Toby Martinez*

Though they did not realize it at the time, the men of Eagle Troop had just fought the main action in what has become one of the most studied battles of modern times, the Battle of 73 Easting. The Army and General Fred Franks (the commander of VII Corps) were so impressed with the results of the fight that a team of analysts from the U.S. Army's Institute for Defense Analysis came out to study every aspect of the battle in order to reconstruct it for future training and use back in the United States. From this has come a computer model of the entire battle, which is considered on a par with textbook operations such as Joshua Lawrence Chamberlain's defense of Little Round Top at Gettysburg and Major John Howard's capture and defense of Pegasus Bridge on D-Day. But before the history books could be written, there was another day of war to finish. For Eagle Troop, though, it was actually fairly quiet.

Tom Clancy: What happened the next day, February 27th, the last day of the ground war?

H. R. McMaster: We had taken some prisoners the previous night—initially forty-two, and then many more sporadically into the next day. We also moved closer to Kuwait.

Tom Clancy: When did you find out about the cease-fire?

H. R. McMaster: The next day.

Tom Clancy: In summary, what were your impressions of the battle you fought?

H. R. McMaster: The Iraqis were unprepared for the American Army. The Republican Guard proved to be a spirited but inept adversary. Our units were much better trained and equipped. I think that the real difference in the fight, however, was the American soldier. Because our soldiers were confi-

General Franks with Captain H. R. McMaster inspecting the battlefield at 73 Easting.
LIEUTENANT COLONEL TOBY MARTINEZ

dent and aggressive, our troop was able to act immediately and as a team. Armored battles in the open desert are decided *very* quickly. We surprised the enemy and were able to take advantage of it. We hit the enemy so hard in the opening moments of the battle, and penetrated his positions so rapidly, that he was unable to recover.

A few weeks after the end of the war, Eagle Troop and the rest of the 2nd ACR packed up and headed back to Germany to resume their normal duties. Sadly, with the end of the Cold War and the success of the U.S. Army in the Persian Gulf, the perceived need for units like the 2nd ACR had decreased to the point where a decision was made to deactivate this distinguished and long-serving unit.

Tom Clancy: What did Eagle Troop do after the cease-fire?

H. R. McMaster: We moved into Kuwait for several days, and later secured a portion of southern Iraq. There we rendered humanitarian assistance to the predominantly Shi'ite Muslim population, and we captured or accepted the surrender of thousands more Iraqi soldiers. Our unit then moved back to KKMC and began to prepare our vehicles for departure back to Al Jubail and eventually Germany. General Franks promised that since we were the first VII Corps unit in, we would be first out. We returned to Bamberg at the end of May.

Tom Clancy: Later on in 1991, you had to hand over Eagle Troop. What did you do after?

H. R. McMaster: It was difficult to leave command. We were notified that the 2nd ACR would be deactivated [in Europe] and re-established in the United States. When Major MacGregor was promoted and left to take command of a cavalry squadron, I became the S-3 [operations officer].

The 2nd ACR moved to Fort Lewis, Washington, and later Fort Polk, Louisiana, reconfigured as a light cavalry regiment. After a few months as S-3, I went to the Combined Armed Services Staff School at Fort Leavenworth, Kansas, after which I began my graduate study at the University of North Carolina, Chapel Hill. In the summer of 1994, I will begin teaching history at West Point.

Tom Clancy: What are your thoughts about your career, after all these experiences?

H. R. McMaster: My professional life has been more satisfying and rewarding than I ever imagined. I would recommend the Army as a career to anyone who has the desire to serve.

Captain H. R. McMaster trained for over a decade to be ready for the relatively short period (probably less than an hour) that he was in intense combat. Yet it is doubtful that he or the taxpayers of the United States would question the bargain they got for their money and efforts. And for all of us, the good news is that H.R. is not all that unique. As the Army Chief of Staff, General Gordon Sullivan, pointed out recently, the U.S. Army is full of fine young soldiers like Captain McMaster.

At the time this book was being completed, H.R. and Katie McMaster got two more examples of how rewarding the life they've chosen can be. Their third daughter, Caragh Elizabeth, was born, and H.R. was selected for promotion. It is good to know that the daughters of H. R. McMaster will know a better world—one that was forged by people like their dad.

Roles and Missions: The ACR in the Real World

A t the end of the Cold War, it was difficult to imagine a crisis large enough to require the Army to deploy a whole division, or even a regiment. In fact, since Vietnam, no U.S. Army unit bigger than a brigade had fought as a unit. Certainly there had been corps-sized exercises, but no situation had actually demonstrated that the Army really needed large-scale units. Some analysts even suggested that the Army should downsize itself to a few brigade-sized task forces.

Desert Storm shattered this theory. The United States fielded and maneuvered three full corps of troops (two Army, one Marine) to action against Iraq. Continuing threats from Iran, Iraq, and North Korea, instability in the Balkans and the former Soviet Union, and the need for large-scale humanitarian and peacekeeping missions, as in Somalia, suggest that the United States needs to be prepared to use the exceptional combat and staying power of ground forces to achieve national objectives.

This said, just how might an armored cavalry regiment be used in the next few years? Let's look at two scenarios that explore the range of options that might be presented to the United States. The first of the scenarios that follow explores the uses of conventional armored cavalry, represented by the current 3rd Armored Cavalry Regiment based at Fort Bliss, Texas. The second scenario looks at a new formation, the Armored Cavalry Regiment-Light, converted from the old 2nd Armored Cavalry Regiment when it returned from its NATO mission. It is a new and untried organization, with many details left to be worked out. Nevertheless, it will probably become a major player in the action to come for the Army's mobile "fire brigades"—the armored cavalry regiments.

Operation Robust Screen: The Second Korean War, January 1997

How they had lasted over fifty years was a mystery. The Democratic People's Republic of Korea was an anachronism—a hermit kingdom from which little information ever escaped, and little of that made sense. But one thing was clear. The North Koreans wanted to control all of Korea. They had fought one war in the early 1950s in a fruitless attempt to gain that goal, and few doubted that they would attack the South again if the opportunity arose. Even though repeated peace overtures from the Republic of Korea (ROK, South Korea) had met with superficial cordiality, they had repeatedly led nowhere. Meanwhile, the spectacular economic development of the South had made the North increasingly irrelevant. Politicians worldwide had adopted the convenient belief that ignoring Kim Il-Sung and his erratic son and designated successor, Kim Jong-Il, would make them disappear. It was a dangerous assumption.

As the third generation of North Koreans grew to maturity, knowing nothing but the two Kims' bizarre blend of militarism, Confucian morality, and Communist dogma, pressures built up within the party and the military elite for a final and forcible reunification of the divided peninsula. For almost fifty years the Inmun Gun (Korean People's Army), a brutally disciplined, lavishly equipped force of over a million, had trained, planned, and prepared for one mission: the "liberation" of the South. Since the end of the first Korean War in 1953, hundreds of thousands of patiently toiling laborers had burrowed into the hard granite mountains of Korea to build underground aircraft hangars, arms factories, command centers, even hardened radar stations with pop-up antennas, protected by massive steel doors. About two dozen of those hundreds of burrows were particularly precious to the Great Leader. They were silos for the home-grown Nodong[1] missiles, with home-grown nuclear warheads.

Even after the death of the elder Kim in 1994, the "Dear Leader" (as Kim Jong II required others to address him) still believed in self-reliance. Like his father before him, the junior Kim wanted to complete the great work of Korean unification as a legacy to the world, before he went to join the other great Communist saints, Marx, Lenin, Stalin, Mao, and his father.

In Korean cosmology, separate seasons and divinities are assigned to each of the four directions. Since north is associated with winter and the Divine Warriors, the Dear Leader thought it fitting that the invasion of the South would begin in January, in the dead of winter—and as luck would have it, just as the American imperialists were inaugurating a corrupt new President. Kim was inspired to compose a poem on the subject, for limited circulation within the ever-appreciative circle of the Central Committee, to celebrate the coming "liberation" of the South. As might be imagined, it was well received.

Sunday, January 25th, 1997, 0300 Hours[2]

To ensure strategic and tactical surprise, the People's Army attacked without advance preparation (having been on winter maneuvers), on less than an hour's notice, under total radio silence, relying on sealed orders. The first wave of invaders included some twenty-two commando brigades composed of over 70,000 elite special forces troops. They swarmed through tunnels under the Demilitarized Zone, parachuted from antique An-2 Colt transport biplanes (quite stealthy because of their wooden construction), or swam ashore from midget submarines. A small team disguised as Japanese businessmen hijacked a Korean Air Lines Boeing 747 in flight, and briefly seized control of Seoul's Kimp'o International Airport, the control tower and terminal complex being thoroughly wrecked when the ROK elite Capital Division stormed it the next day. One of the most successful of the North Korean special forces brigades crash-landed inside the U.S. Embassy compound, using a number of American-made MD-500 helicopters illegally acquired from a German arms dealer back in the early 1980s. The Marine guards were wiped out, and the handful of Embassy staff then on night duty were slaughtered. When a hastily assembled relief force of U.S. Army Military Police and combat engineers arrived to retake the building, it was in flames, with its vital electronic communications and monitoring gear destroyed. A similar raid on U.S. Eighth Army[3] headquarters in the suburb of Yongsan was detected in time and decimated by the Stinger missiles of an alert Avenger air defense battery.

Sunday, January 25th, 1997, 1200 Hours

Most of the North Korean commandos were wiped out quickly, but the confusion and disruption they spread helped to open the way for the main attack. The rugged topography of Korea allows for only a few invasion routes, and these tend to channel the flow of any military movements. The narrow road down the east coast barely provided maneuver room for a single North Korean division of the 806 Mechanized Corps, which seemed to pay for every yard with a knocked-out tank. Five specialized river-crossing regiments and several divisions of infantry forced the wide Imjin River along the west coast, but the bridgeheads were contained and gradually eliminated by the ROK divisions holding the line. The main axis of advance was the highways east of Seoul. There were 2,000 T-72, T-62, and improved T-55 tanks, supported by more than a dozen tube-artillery regiments and some sixty-plus rocket-artillery battalions, massed along a front of less than 50 miles/82 kilometers in width. The U.S. Army's 2nd Infantry Division, supported by Republic of Korea (ROK) units on either flank, fell back toward Seoul taking heavy losses, but inflicting three or four times as much damage as it was suffering. The People's Army knew that street fighting always favors the defender, and the Dear Leader wanted to take the historic, economic, and cultural center of the nation relatively intact. The invasion pushed south and east, away from the heavily urbanized capital area and down the Han River Valley. The North Korean objective was to bypass Seoul, then hook suddenly westward to capture the ancient walled town of Suwon. The capital with its ten million residents would be cut off, besieged, and starved into surrender. What the Dear Leader's generals did not know was that this was exactly what the the Eighth Army wanted them to try. The North Korean army took the bait.

Despite the chaos of the war's opening hours, the situation was quickly assessed by the National Military Command Center in the Pentagon, and the Pacific Command (PACOM)[4] headquarters in Hawaii. North Korean frogmen had cut the telephone and fiber-optic cables that crossed the Tsushima Strait to Japan, and relentless rocket and artillery attacks forced the surviving headquarters units in South Korea to move constantly, making it difficult to maintain communications even via satellite link. But real-time imagery pouring in from reconnaissance satellites in low earth orbit made it abundantly clear that the Second Korean War had begun. The new President, who had taken office just a few days before, was informed by a call from the Chairman of the Joint Chiefs of Staff, an Air Force general. The President immediately convened a meeting of the National Security Council, and asked the Speaker of the House to arrange an emergency joint session of Congress. Meanwhile, as Commander in Chief, he ordered the acting Secretary of Defense to execute the existing plans for the reinforcement of Korea. Less than one hour later, the officer on duty in the communications cell of the U.S. III Corps headquarters at Fort Hood, Texas, received an urgent phone call.

Monday, January 26th, 1997, 1000 Hours

The first reinforcement unit to arrive in Korea was the Alert Brigade from 82nd Airborne Division. Airlifted directly from Fort Bragg, North Carolina, to Taejon, Korea (a 7,000-mile flight that took almost twenty hours with refueling stops), they deployed the next day into the hills north and west of the city to secure the air base and the strategic bridges over the Kum River. With the airfields around Seoul under continuous SCUD and long-range artillery bombardment, Taejon was chosen as the forward headquarters of the U.S. IX Corps, based in Japan, which would control most of the units being sent to reinforce the Eighth Army.

At the same time, six Maritime Prepositioning Ships (MPS)[5] steamed out of Agana Harbor, Guam, bound for Pusan with supplies and equipment for a Marine brigade. Except for one quick-reaction battalion aboard several amphibious transports at Okinawa, the troops would fly in from Camp Pendleton, California. Thus, the initial mission of the 1st Marine Expeditionary Force was to secure the ports of Pusan and Ulsan and keep them open. Once this was certain, the leathernecks would move up to the line wherever the North Korean threat was greatest, and dig in.

At the same time a second MPS squadron left Guam with equipment for a brigade of the 10th Mountain Division (Fort Drum, New York). The troops would be airlifted into Taejon by the end of the week and rushed north to relieve the battered 2nd Infantry Division, which would be pulled back into the Seoul pocket for reorganization and a bit of rest. As the C-5 Galaxy, C-17 Globemaster III, and Civil Reserve Air Fleet (CRAF) transports returned from delivering the first wave of reinforcements, the alert brigade from the 101st Air Assault Division (Fort Campbell, Kentucky) would be airlifted to Taejon to form an airmobile reserve with enough helicopters to move the entire brigade in a single lift.

The linchpin of the reinforcement plan was the 3rd Armored Cavalry Regiment (3rd ACR) at Fort Bliss, Texas. With 123 M1A2 Abrams tanks, 127 M3A2 Bradley Fighting Vehicles, seventy-four helicopters of various types, and hundreds of wheeled vehicles, the regiment was impractical to airlift, especially with the extra engineer, artillery, military police, and support battalions attached from the III Corps at Fort Hood[6] Even if there had been enough transport aircraft (and cutbacks in the procurement of the C-17 during the 1990s meant there weren't), the few remaining operable airfields in Korea were overwhelmed with arriving supplies, reinforcements, and casualty evacuation. During the next few weeks, runways and terminals were also under sporadic attack from rockets and mortars carried by North Korean infiltrators. Despite the claims of airpower enthusiasts, you don't just airmail an armored unit the way you would an overnight letter. The 3rd ACR would have to go by boat. But first, it had to get to the boats, and that was a trick in itself.

Wednesday, January 28th, 1997

You need specially reinforced rail cars to transport 70-ton tanks like the M1A2. Yet thanks to tireless staff work by planners and dispatchers at the Military Transportation Command in St. Louis, Missouri, it took just over two days to gather the rolling stock from across the country and assemble complete trains on the rail sidings at Fort Bliss and Fort Hood. Meanwhile, the Military Sealift Command sent six SL-7 roll-on/roll-off cargo vessels to Long Beach, California, and the two additional SL-7s (there are only eight of them) to Beaumont, Texas.

The SL-7s are vessels with a remarkable history. Built in the early 1970s by German and Dutch shipyards as very large high-speed container ships, they were too expensive to operate and maintain commercially. But the combination of speeds of 30+ knots and immense capacity was irresistible to a Sealift Command that had seen its WWII-era cargo ships decay into obsolete hulks while the U.S. Merchant Marine withered away. Displacing about 30,000 tons empty and 55,000 tons at full load, an SL-7 can hold 180 heavy tanks, or 600 HMMWVs. Each SL-7 has a pair of 50-ton cranes, as well as roll-on/roll-off ramps port and starboard. The ships were named for eight navigational stars long special to mariners: *Algol, Bellatrix, Denebola, Pollux, Altair, Regulus, Capella,* and *Antares.*[7]

Following plans that had been carefully worked out in countless exercises and simulations, the 3rd ACR was loaded at Fort Bliss, two trains a day, while the III Corps loaded one train every two days at Fort Hood. As each element of the regiment arrived in Long Beach, California, it was loaded onto its designated SL-7. The same process was repeated at the port of Beaumont, Texas, for the III Corps artillery and other attached units.

The transportation plan was based on the concept of "combat loading." This means that every vehicle would be fully fueled and armed when it rolled off onto the docks in Pusan. This made the ships more vulnerable to fire and explosion if they were hit, but it reduced the time required for the 3rd ACR to prepare for battle upon arrival. Critical regimental assets were carefully divided up among different ships, so that the loss of a single vessel would not cripple the regiment. The Abrams tanks were driven onboard the ships already loaded on Heavy Equipment Transporters (HETs). This took up more space, but ensured that the armor could rush to the front at high speed over the excellent South Korean highway net, without wear and tear on tracks or suspensions.

Monday, February 9th, 1997

The SL-7s from Long Beach took six days to cross the Pacific. The ships from Beaumont had to pass through the Panama Canal, adding about three days to the trip. Meticulous work by intelligence and Special Operations troops attached to U.S. Army Southern Command Headquarters at Fort Clayton, Panama, identified and "terminated" a North Korean sabotage team traveling on fake Chinese passports that had been dispatched to sink a rusty old Panamanian-flagged cargo ship in the canal's narrowest section, the Galliard Cut. For February, the weather was unusually mild in the stormy North Pacific, and knowing the urgency of the situation, the civilian (but mostly ex-Navy) crews of the big ships extracted every bit of performance from their temperamental boilers and steam turbines. Thus the little convoy averaged a bit more than the specified thirty knots.

As they rounded the southern tip of the Japanese home islands and entered the Korea Strait, the convoy was met by an escort of four O.H. Perry-class frigates. Hastily mobilized from the Navy Reserve Force, they were the only available ships with medium-frequency sonars suitable for detecting enemy submarines in those shallow waters. This proved to be a wise precaution; as the convoy approached Pusan, a wolfpack of obsolescent North Korean Romeo-class submarines, lurking off Tsushima Island, was detected and annihilated by torpedoes dropped from the frigates' helicopters before the subs could close to attack range.

Tuesday, February 10th, 1997

When the ships docked at Pusan's magnificent north harbor, the first units to off-load were the 27 MLRS launchers of the 6th Battalion of the 27th Field Artillery Brigade. Back at Fort Hood, each vehicle had been loaded with a pair of ATACMS, a chubby guided missile with a 60-to-90 mile/100-to-150 kilometer range. The drivers and gunners, who had flown in the previous night, were already waiting on the docks to take delivery. The launchers drove off the pier and onto rail cars headed north. Next to disembark were the fifty-two attack and scout helicopters of the 3rd ACR's 4th Squadron. They were immediately flown to Pyongtaek airfield, forty miles south of Seoul, where advance teams had set up a Forward Arming and Refueling Point (FARP).

Wednesday, February 11th, 1997, 0700 Hours

The next morning, pairs of AH-64A Apache and OH-58D Kiowa Warrior helicopters fanned out across the hills, east and west of the broad Nam Han River Valley, where the main enemy thrust southward was developing. The Kiowas, with their mast-mounted laser designators and thermal sights, could peek over the ridgelines, spot a target, and call in a supersonic Hellfire missile launched from an Apache flying miles away, concealed beyond the next line of hills. The air cav gunners concentrated on anti-aircraft systems, particularly the old but deadly S-60 towed 57mm guns and the new armored scout vehicles carrying twelve-round SA-18 missile launchers.

As the regiment's three other cavalry squadrons unloaded and rushed north along the Seoul-Pusan expressway, the 3rd ACR was assigned a "fire brigade" role, to plug gaps in the line and stop any enemy spearheads that broke through the ROK's determined defense.

Highway 327 crosses the Han River near the village of Punwon-ni; ROK engineers had blown the bridge as soon as enemy recon units approached the north bank. The fresh North Korean 820th Armored Corps and 815th Mechanized Corps, backed by a division of artillery, were ordered to cross the river in this sector and secure a bridgehead on the south bank. Though they were traveling only at night—without lights and using superb camouflage discipline to hide out from satellite reconnaissance during daylight—the enemy movement was still observed and tracked by the aero scouts of the 3rd ACR and reported to the forward command post of Colonel Rodriguez (the 3rd ACR commander) near Suwon.

The river line was held by a division of ROK reservists that had been badly chewed up in the withdrawal from the DMZ two weeks earlier, losing most of their vehicles and heavy weapons. But they still had their entrenching tools, M16s, and a dwindling supply of TOW and Javelin anti-tank weapons. The colonel commanding the ROKs (both of the generals had been killed in action) knew his country had no more space to trade for time, and his men were determined to hold the riverbank or die in place. They were under constant bombardment by whole brigades of rocket launchers, heavy mortars, and field guns.

Meanwhile, in the low hills northwest of the crossing point, the enemy was assembling a river-crossing force, including engineers with mobile pontoon bridging equipment, a regiment of light amphibious tanks, and a brigade of commandos with inflatable assault boats. This far south there were hardly any ice floes in the river. The NK corps commander had trained these men for years under far worse conditions. He might drown half of them in the frigid waters of the Han, but he *would* get a foothold on the south bank. Then he would push his reserve division across, surrounding the ROK puppets of the U.S. imperialist aggressors and opening the road to liberate Suwon. After that, he could wheel south and drive the rest of the Americans and their Korean lackeys into the sea. He imagined his T-72 command tank would be the first unit to make a triumphal entry into Pusan.

Thursday, February 12th, 1997, 0100 Hours

Colonel Rodriguez paged down through the weather forecasts on the high-resolution color LCD screen of the Silicon Graphics BattleSpace

Workstation in his M4 command track. The next morning would be foggy in the lower Han Valley, and the fog would not lift until midday. He smiled as his fingers danced across the keyboard, instantly transmitting orders over the secure satellite data link to his squadron commanders and attached combat-support units. He could have dictated the words to one of the three enlisted console operators, but everyone knew he was the fastest computer jock in the regiment, a holdover from his days at West Point.

Back at the 4th Squadron FARP, CW-3 (Chief Warrant Officer Third Class) Jennifer Grayson worked her way through the short preflight checklist for her OH-58D helicopter. She had been through this drill 376 times, but she never took shortcuts or skipped a step. There were still some Neanderthals in the Army who thought that a woman shouldn't be a combat helicopter pilot; thus she had always striven for "zero defects." Her copilot, WO-1 (Warrant Officer First Class) Greg Olshanski, loomed up out of the predawn darkness carrying the DTD (Data Transfer Device), a little gadget resembling a video game cartridge. He inserted it into a socket on the crowded instrument panel, automatically loading the mission's assigned radio frequencies, navigational waypoints, and IFF mode codes. The DTD would remain in its socket recording critical flight data from the Kiowa's control system, for after-action review. A blank videotape was already loaded in the helicopter's onboard video cassette recorder to capture a permanent record of every target engagement. "Our call sign tonight is Nomad Two-Seven," said Olshanski.

"Nomad Two-Seven," CW-3 Grayson grunted in acknowledgment. The immediate threat to the river line was enemy armor, so the Kiowa was loaded for tank-busting, with four Hellfire missiles on the weapons pylons. Grayson missed having the .50-caliber machine-gun pod—Hellfires were too easy—and she liked to shoot up trucks and soft targets with the .50. That took some skill, and a light touch on the controls. She had both.

Thursday, February 12th, 1997, 0400 Hours

The imaging infrared camera on a stealth recon drone sent out during the night by the IX Corps intelligence battalion had spotted an enemy armor battalion of thirty-one tanks moving down Highway 327. With the approach of daylight, they had pulled off the road and dispersed into a narrow canyon. Grayson pulled up the thermal view on her multi-function display. The tank engines would still be warm by the time the OH-58D came into range. The North Koreans were good at camouflaging their tanks with netting, tree branches, and shrubbery; but the rear decks of those T-72s would stick out like sore thumbs to the thermal viewer in the mast-mounted sight.

Grayson and Olshanski carefully timed their arrival at each waypoint. There was a lot of traffic in the air this morning, and most of it was flying without navigation lights or search radar to give away its position. Some of the traffic consisted of artillery shells, blindly obeying the laws of physics. Air cav planning staffs devoted a lot of effort to "deconfliction"

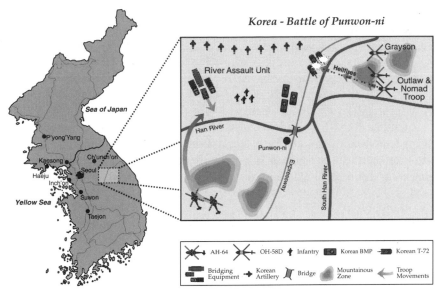

Korea - Battle of Punwon-ni

The battle of Punwon-ni. Helicopters from the 4th (Air Cavalry) Squadron of the 3rd Armored Cavalry Regiment blunt an attempted river crossing of the Han River by the North Koreans.

JACK RYAN ENTERPRISES, LTD., BY LAURA ALPHER

with their field artillery counterparts, making very, very sure that friendly helicopters and friendly projectiles never tried to share the same airspace at the same instant.

Grayson steered the agile chopper up behind the crest of a mountain spur. The enemy tanks lay just across the ridge, their crews already bedded down, except for a few sentries nervously scanning the skyline. With a delicate nudge of the cyclic and a gentle adjustment of the collective, she rose a few feet, so that the spherical head of the mast-mounted sight, like the face of a grotesque three-eyed robot, peered over the rocky lip of the valley. She flicked the arming switches on the Hellfire control panel, aimed, and fired. Reflexively, she closed her eyes for a second, so that her night vision would not be dazzled by the flash of the rocket motor as it came off the rail, rose in a graceful arc, and dropped directly onto a tank 2,000 yards away. Before the first round struck, the next was on the way. Then another. Within a few seconds three tanks had exploded, the lethal mixture of diesel fuel and ammunition blowing the turrets completely off the vehicles. Within a few more seconds, the startled North Korean crews of a dozen tanks had recovered and were directing bright tracer streams of 14.5mm machine-gun fire at the hilltop. But the helicopter was already hidden behind the ridgeline, calling over the Automated Target Handoff System (ATHS) for other helicopters to join in the carnage. With three missiles expended, the OH-58D was four hundred pounds lighter and would have tended to rise into full view of the alerted enemy. But as Grayson swiftly and instinctively compensated for the weight change, the chopper swooped down and to the left, evading the return fire.

A voice crackled over the radio headset, "Nomad Two-Seven, this is Outlaw Four-Six, I'm about two clicks behind you with sixteen rounds. What have you got for me? Over."

"Roger that, Outlaw Four-Six, this is Nomad Two-Seven. We have two dozen Tango Seven-Twos at our ten o'clock, approximately two clicks out. They're pretty stirred up right now. We can start designating targets for you in thirty seconds. Go to Mission Package Alpha Seven, over."

Outlaw Four-Six was an AH-64A Apache, with a full load of missiles and a 30mm automatic cannon. The two crews set all the necessary switches for an automatic handoff from the ATHS. The OH-58D would play hide-and-seek around the rim of valley, designating targets with its laser while the AH-64 stood off at a safe distance firing missiles. The first missile was already in flight toward an unlucky T-72 when the voice of Lieutenant Colonel Martin, 4th Squadron commander, broke in on the squadron command net. All units of Outlaw and Nomad troops were ordered to abort their current missions and close as rapidly as possible on a new set of target coordinates some miles to the west. An ROK scout platoon had spotted the enemy river-crossing task force moving toward the northern bank of the Han.

"This is going to be hairy. I wish we had the .50-cal," said Olshanski.

"This is what we get paid for," Grayson replied grimly, punching the new coordinates into the navigation system. To reach the assembly area where the North Koreans were preparing to force a river crossing, the air cav squadron had to run a gauntlet of small-arms fire and shoulder-launched SA-18 missiles. (Actually they were North Korean copies of the Chinese copy of the Russian SA-18. They weren't very reliable, but there were lots of them in the air.) Flying low and dodging constantly, Grayson reached the target area and saw long columns of boxy shapes waddling down toward the riverbank through the MMS FLIR system. There were PMP pontoon bridge sections, GSP tracked self-propelled ferries, and PTS-M tracked amphibious transporters. The North Koreans had acquired (at bargain-basement prices) some of the vast menagerie of river-crossing equipment the Soviets had designed to cross the Elbe, the Rhine, the Moselle, and the Meuse (see *Red Storm Rising* to recap this Cold War scenario). Rugged and cleverly engineered, these vehicles had come a long way to cross this river. Grayson intended to make sure their journey had been in vain.

There was a low ridge a hundred yards back from the south bank of the river. Grayson and a few other OH-58Ds swung in behind the ridge and began popping up to designate targets for the Apaches, which found safe firing positions a mile or two further back. The North Koreans had dug in a few batteries of ZU-23 twin 23mm anti-aircraft guns to cover the crossing site. These were first-priority targets for the Hellfire missiles. Then the columns of bridging equipment and truckloads of assault boats were raked with missiles, creating a huge smoking, burning traffic jam for almost a mile back from the river. The Apaches now closed in to complete the destruction with salvoes of unguided 2.75"/70mm rockets and bursts of 30mm cannon fire.

Off to her left, Grayson saw a flash and a puff of dark smoke. A North Korean SA-18 struck Outlaw Four-Three squarely on the tail boom, shredding the tail rotor. The Apache spun out of control toward the frozen ground on the enemy side of the river. Fortunately, the helicopter was flying low enough that the crash looked survivable. Grayson clicked the radio transmitter to the Squadron net frequency. "This is Nomad Two-Seven. Cover me, I'm going in to pick them up, over," she said.

A live American helicopter crew was a prize worth taking risks for. Senior Sergeant Kim Cho-buk was a twice-decorated Hero of Socialist Struggle, a First Class Heavy Machine Gun Marksman, and acting commander of an armored reconnaissance platoon (after the Lieutenant's BRDM scout vehicle had taken a Hellfire missile through the roof that morning). The gunsight of his one-man turret was crude; but at this range, it took little marksmanship to pour a stream of bullets into the falling Apache as it slammed into the riverbank. Kim kicked his driver between the shoulder blades and screamed at him to close in. The other BRDM in the platoon followed a hundred meters behind; and some infantry squads nearby rose from their foxholes and started running toward the downed aircraft (probably hoping the Americans had some MREs on board).

Jennifer saw two enemy scout vehicles and some running dismounts break out of cover and head toward the crash site. She saw a stream of tracers as the lead scout vehicle fired. She barely noticed as Olshanski nailed the BRDM with their last Hellfire. She was concentrating on keeping a low stable hover as close as possible to the wreck, where two dazed and bleeding aviators were struggling out of their harnesses.

Outlaw Four-One, another Apache, rolled in a few hundred meters behind Grayson. As it opened up with the 30mm cannon, the ragged line of North Korean infantry fell back, and the scout vehicle popped smoke grenades and slammed into reverse gear.

The crash survivors staggered over to the hovering OH-58D and hooked their harnesses onto the landing skids. It looked crazy but it was a standard operating procedure for combat rescue. As she lifted off with two windblown but very grateful warrant officers dangling securely from the skids, Jennifer still wished she had a .50-cal on board.

The Punwon-ni sector of the river line held, but that night the North Koreans secured a bridgehead further downstream, got a mechanized corps across, and pushed south to cut the expressway at Pangyo-ri, between Seoul and Suwon. If they could take Suwon and drive through to the west coast, the Seoul-Inchon metropolitan area would be cut off, with 40% of the nation's people and most of its economic might.

Friday, February 13th, 1997, 0630 Hours

Aero scouts and ground-based recon units carefully pinpointed the North Korean artillery positions and command posts of the enemy divisions converging on Suwon. Just before dawn, three battalions of MLRS deployed back in Taejon fired a salvo of ATACMS. As the warheads detonated high above the battlefield, they rained cluster munitions over an area of several square miles. Virtually the only survivors were inside armored vehicles or dug in underground. The morning fog still lingered in patches over the frozen rice paddies, when 2nd and 3rd Squadrons of the 3rd ACR broke out of the foothills and tore into the flank of the North Korean 678th Mechanized Rifle Division. The M2A2 Bradley cavalry vehicles found good hull-down firing positions behind the earth embankments that separated the fields. As they picked off enemy command vehicles (conspicuous because of their extra antennas) with long-range TOW missile shots, the tanks swept forward at high speed, firing on the move at anything that fired back. Anti-tank rounds from dug-in 122mm guns glanced off the M1A2 turrets and front plates as if they had been fired by peashooters.

A few North Korean anti-tank teams popped out of concealed foxholes to fire after the tanks passed, disabling several M1A2s with wire-guided missile shots into the thinly armored rear engine compartment. Before they could get off a second shot, most of the missile teams were spotted and cut down by machine-gun fire from the Bradleys. Meanwhile, one tank in each platoon had been fitted with a hastily improvised dozer blade to slice through the rice paddy embankments (the original supply having been lost in a freak SCUD hit back at Pusan). A welder in the regiment's 43rd Engineer Company had seen pictures of the "hedgerow cutters" fitted to M4 Sherman tanks in Normandy during 1944, and had thought he could improve on the idea. His captain had taken the idea to Colonel Rodriguez, who had immediately approved it. Welders don't usually get medals, but this one would. The tankers appreciated the immediate improvement in their cross-country mobility. There's an old saying that "speed is armor." Now they had both.

Sunday, March 1st, 1997

The battle at Pangyo-ri proved to be the high-water mark of the North Korean invasion. Over the next three weeks the front stabilized along a track running from Sokcho on the east coast, through the rubble of Ch'unch'on, and down the northern Han River line to the outskirts of Seoul.

After suffering 50% loss rates in furious air-to-air battles during the war's first week, the North Korean Air Force kept its surviving MiGs in their rock tunnel shelters, conceding air superiority to the Americans. U.S. Air Force B-1s by day, and F-117As (and even a handful of B-2s) by night, kept up a steady offensive against enemy supply lines, command centers, and artillery positions. Occasional SCUD missiles caused damage and civilian casualties in the South Korean cities, but they could not stem the constant flow of fresh units and supplies. More important, the balance of terror held—the Dear Leader was not crazy enough to unleash the nuclear, chemical, and biological holocaust that slept silently in his deepest underground bunkers.

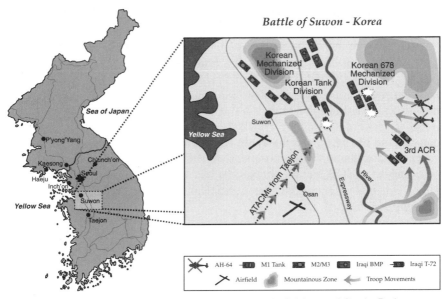

Battle of Suwon - Korea

The battle of Suwon. Supported by ATACMS missile strikes, the 3rd Armored Cavalry Regiment conducts a spoiling attack to stop the North Korean drive to the Yellow Sea.

Jack Ryan Enterprises, Ltd., by Laura Alpher

The U.S. 1st Cavalry Division (the First Team), from Fort Hood, Texas, and the 1st Mechanized Infantry Division (the Big Red One), from Fort Riley, Kansas, began disembarking in Pusan during the first week of March to provide Eighth Army with an offensive option.

Meanwhile, the Dear Leader had contemptuously ignored so many UN resolutions that on March 13 the DPRK became the first nation ever expelled by the General Assembly.

The cherry blossoms were already blooming on hillsides that had not been ravaged by shell fire when Eighth Army struck back. The entire U.S. 1st Marine Expeditionary Force[8], quietly joined by a British Royal Marine battalion and a brigade-sized task force of French light armor, had boarded amphibious assault ships and steamed into the Yellow Sea, escorted by a battle group built around the carriers *Constellation* and *Theodore Roosevelt,* threatening the long west coast of the peninsula. In consequence, the North Koreans had to tie down a dozen infantry divisions in static coastal-defense missions. They expected a replay of Douglas MacArthur's surprise 1950 Inchon landing somewhere along their long and vulnerable coastline. They were fooled.

Tuesday, March 31th, 1997, 0530 Hours

3rd ACR, heavily reinforced with artillery, engineer, and reconnaissance units, led the IX Corps assault north and west out of Chonpyongchon, with the armor-heavy U.S. 1st Cavalry Division close behind. Critically short of fuel to maneuver, the North Koreans could do little but dig in and wait to be bombarded, cut off, surrounded, and bypassed. On the first day, the cavalry squadrons advanced over twenty miles, while the air cav squadron ranged forty or fifty miles deeper to shoot up supply trucks and rear-area headquarters units. On April Fool's Day, the old 2nd Infantry Division base at Tongduchon (Camp Casey) was recaptured in bitter fighting; and elements of twelve enemy divisions were trapped in a pocket around Uijongbu. As the enemy's air-defense missile and ammunition supply ran out, the Marines were brought ashore by relays of helicopters, in deep "vertical envelopments." Units began to surrender—instead of fighting to the death—by squads and platoons on the second day, by companies and battalions on the fourth. Less than a week after the start of the counteroffensive, the advance of the cavalry squadrons reached the DMZ, brushed aside weak resistance at Panmunjom, and took the town of Kaesong inside North Korea.

Wednesday, April 15th, 1997, 1200 Hours

The North Korean situation clearly was hopeless. It was no surprise when the noon broadcast from Pyongyang Radio announced that the Dear Leader, and the top leaders of the Workers' Party had been taken into custody, and the provisional military government was requesting a cease-fire and immediate negotiations for reunification of Korea. It was April 15th, but the taxpaying citizen soldiers of the Eighth Army felt that this time, they had gotten their money's worth.

Operation Rapid Saber: Uganda, June 1999

The 2nd Armored Cavalry Regiment (2nd ACR, The Dragoons) had been reconfigured in the early 1990s as an Armored Cavalry Regiment-Light (ACR-L), an easily transportable armored unit to provide mobile armored firepower to the troops of the XVIII Airborne Corps, typically the first American soldiers to deploy when an emergency is too far from the shore for the United States Marines. The Army fought long and hard to have this unique unit equipped with the latest technology. Designating it an experimental unit helped (for Pentagon accounting purposes), though its performance in maneuvers was the best justification for the expense. The M1 Abrams tanks had been replaced one-for-one with the new M8 Armored Gun System (AGS). In addition, all of the Bradleys had been replaced with M1071 "Heavy Hummers"—HMMWVs protected by advanced composite armor. Every vehicle was "wired" into the IVIS command and control network. Some were equipped with .50-caliber machine guns, others carried Mk-19 40mm grenade launchers and lightweight TOW launchers. About one out of every five carried a new weapons system, the Non-Line of Sight (N-LOS) missile, with eight missile rounds in a vertical launcher. Every 2nd ACR-L dismounted trooper had the new Virtual Battlefield helmet, with a built-in

GPS receiver, helmet sight, and data link onto the IVIS network. They called themselves "Starship Troopers."

As part of the XVIII Airborne Corps stand-alert force, one of the regiment's three armored cavalry squadrons was always kept on alert, and a wing of Transport Command (TRANSCOM) airlifters was similarly kept "hot" to transport the squadron. It was mere coincidence that when the Uganda Crisis broke, the 2nd Squadron of the 2nd ACR-L had "the duty," along with the 512th Military Airlift Wing, a reserve outfit at Dover, Delaware, which would be the first off the flight line to Fort Polk, Louisiana.

Uganda, June 1999

Nobody ever expected Idi Amin to reappear on the world stage. Since he was thought to be terminally ill from venereal disease (or already long dead), his return to Uganda was as unexpected as the Israeli rescue mission to Entebbe in July of 1976. With the help of Sudanese and Libyan agents, he escaped from his maximum security (but luxurious) house arrest in Saudi Arabia. Then, with the help of Sudanese "volunteers," he scattered a handful of demoralized border guards and swept into Kampala, the capital city of Uganda. The self-proclaimed "Field Marshal" and "President for Life" and his armed followers quickly took control of the airport, the TV and radio station, the central bank, and as many of the country's fourteen million emaciated citizens as they could abuse and bully. Though ravaged by disease and chronic anarchy, the tragically unlucky nation in the Central African Highlands had nevertheless begun to recover in the late 1990s. There had been enough security and order to allow the return of the UN AIDS Task Force. The international team of 200 physicians and nurses had been in-country for a mere five weeks, with some promising new treatments for the lethal virus that had infected over half the Ugandan population. Amin's first official act was to seize the medical personnel and demand, as a condition of their release, international recognition of his return to power. The murder of the mission leader, a French physician from Pasteur Institute whose resistance to Amin's thugs had been just a little too courageous, immediately crystallized the nature of the crisis. As the bloody pictures appeared on televisions worldwide, phones were picked up, and pre-set contingency orders activated. It was the French President who uttered the words that set things in action, though his choice of words jolted the American chief executive:

"No peace with Bonaparte."

For the French Republic, the killing of a French citizen was a matter of honor, and the Force Reaction Rapide (FRR) began to form up. But the French light-infantry force was indeed light, with little more than machine guns and a 30mm automatic cannon on their lightly armored scout cars and some shoulder-fired anti-tank missiles. Talking heads all over the world noted this, and pointed to aging but quite real Russian T-72 tanks, Mi-24

Hind helicopters, and MiG-29 fighters visible on the Russian real-time satellite reconnaissance photographs of Kampala and Entebbe now available to CNN and other news media. Crewed mainly by Libyan and Sudanese "volunteers," and reinforced by loyal survivors of the old Ugandan Army (recruited mainly from Amin's own small Kakwa tribe), the force was organized into three ragtag brigades and an air wing. Amin recruited enough Egyptian and Pakistani renegade mercenary technicians to keep the engines tuned and the radars calibrated. To prove his sincerity as a champion of militant Muslim fundamentalism, Amin began the systematic massacre of Ugandan Christians. That played well in Khartoum and Benghazi, and kept the money and ammunition flowing.

No match for a Western division—indeed, no match for the armor-heavy 3rd ACR—Amin's army was enough to outgun the French FRR and render a rescue of the international medical team impossible—so agreed the talking heads on news-analysis shows worldwide. None of the talking heads, however, had ever met General du Brigade Jean-Jacques Beaufre or Lieutenant Colonel Mike O'Connor, a Legionnaire and a cavalryman, both veterans of Desert Storm. Professional soldiers hate doing things on short notice. When human life is at stake, careful planning is the minimum requirement, but the danger to civilian lives in this crisis precluded normal concern for the lives of soldiers. That was part of the job, too.

The multi-national action group for the operation was as curious as the mission planners. Intelligence came from overhead imagery developed commercially from Russian recon satellites under contract to Agence France-Presse. The French would be first on the ground and needed the data the most. There was an agreeably flat spot fifty kilometers west of the objective. It was—had to be—close enough, because the Ugandan Army still remembered what had happened when the Israelis made their unexpected visit to Entebbe. All three runways at Entebbe, and the connecting taxiways, were solidly blocked by lines of parked trucks, tanks, and armored vehicles. To discourage helicopter assault, 23mm anti-aircraft guns and surface-to-air missiles were dug in all around the airport perimeter. Weather information came from NOAA, and looked good during the probable operations time "window." The mild climate of Uganda presented few problems, but the clouds of mosquitoes that rose from the lakeside marshes every evening made malaria precautions essential.

The U.S. Defense Mapping Agency has the best cartographic information in the world, and gigabytes of it started flowing over a satellite data link to Paris. All the while, secure phone lines burned between two frantic operations staffs, laboring to do the impossible in two languages at once. With more than a little screaming and profanity (thankfully not fully understood by either side), an operational concept was rapidly hammered out, just as the American forces assigned to the operation boarded their C-17 and C-5 transports for the long hop to Africa. While the world watched replays of the French physician's death on CNN, and the talking heads worried about the

Uganda - Operation RAPID SABER

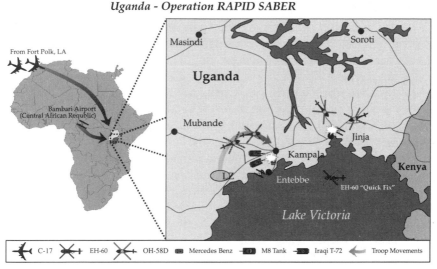

Maps of the approach by Allied forces on Uganda. C-17s from Fort Polk, Louisiana, arrive with elements of the 2nd Armored Cavalry Regiment-Light, as well as French special-forces personnel.

JACK RYAN ENTERPRISES, LTD., BY LAURA ALPHER

expected negotiation process, Operation Rapid Saber, the first-ever airborne armored cavalry mission, got under way.

The discussion of the plan at the Elysee Palace was brief. General Beaufre's force included an elite volunteer hostage-rescue team, detailed from the Direction General de Securite, wearing Foreign Legion paratroop uniforms and carrying fake identity papers. Though they were a critical national asset that the Minister of Defense was reluctant to risk, the president told him the honor of France was at stake. When the Minister of Defense was about to raise an elegantly worded, logical, and tactful objection, the French president slammed down his bottle of Perrier on the antique, polished, inlaid surface of the cabinet table and told him to shut up and have the men at the airfield in two hours.

June 23, 1999, 0400 Hours[9]

The hardest mission fell to the tanker aircraft, decidedly unglamorous birds, mainly flown by Air Force Reserve crews—most of them plucked from their airline jobs—so rapidly called into service that FAA rules for crew rest time on domestic airlines were quietly violated for the next several weeks. One by one, the cargo transports linked up with the aging KC-135Rs and newer KC-10As, topping off their fuel tanks. Distance, time, and tanker assets dictated a direct great-circle course that overflew several African countries. Fortunately, most of these were former French colonies, and through a combination of quiet diplomacy and well-placed French nationals in the various air-traffic-control centers, the 300-mile-long stream of American aircraft flew the width of Africa as uneventfully as a red-eye flight from LAX to JFK.

June 23, 1999, 1930 Hours

The French paratroopers arrived first, just eighty minutes before the American aircraft. Two battalions landed just after sunset, worried that the sun was glinting off the contrails of the transports—but it was too late for that. On hitting the ground, the squads assembled and formed up into platoons that raced outward to secure a perimeter. Crisp, short, radio transmissions reported negative contact with hostile forces. A few bewildered civilians were located and held. Three telephone lines were found and cut, along with every electrical power line shown on the satellite photos. The landing zone instantly became a black hole of information. Anyone unlucky enough to wander in was quickly captured and held in a detention area for later release.

Satisfied that the landing zone (LZ) was secure, the senior French officer got on his radio and called in the approaching transports even before his troopers deployed the lines of chemical landing lights on the hard red dirt of Central Africa. First in was the French command group. General Beaufre quickly set up his CP at a pre-arranged point, while the other transport aircraft shuttled in and out. Not even taking the time to stop their engines, they lifted off quickly to refuel at Bambari in the Central African Republic. Beaufre now had just under a thousand elite paratroopers on the ground, but armed with only light weapons and a few Renault jeeps. Their commanding general grumbled about his country's meager airlift capacity, a mere four squadrons with some seventy aging C-160 Transalls. Well, maybe they'd listen to him next year....

June 23, 1999, 2050 Hours

Lieutenant Colonel Mike O'Connor, sitting in the jump seat of the lead C-17 transport, was thinking the same thing. Watching the approach through night-vision goggles, he caught the glow of the chem lights, while a twenty-nine-year-old Air Force captain named Tish Weaver flared her aircraft in for a soft landing. The pilot's acute senses felt the impact and pronounced it good. The ground here was firm enough for the rest of the first "stick" of transports to land safely, and probably even for the older C-5s, known derisively as Freds [F**king Ridiculous Expensive Disasters] to the drivers of the newer C-17s. First off was Lieutenant Colonel O'Connor in his command HMMWV. While his driver negotiated his way to the CP, two radio operators rigged their antennas, and a traffic-control officer formed up with his French counterparts. When O'Connor reached the CP, salutes were exchanged, and the two unit commanders sized each other up face-to-face, instead of via picture phone and radio. Together they huddled over maps during the two hours it took for the long line of jet transports to land and rig their loading ramps. They had one more hour to assemble, give final briefings, and move out. Every soldier for miles around cringed from the roar of fan-jet engines. Nobody could believe that the mission was still covert. If the MiG jets and Hind-D attack choppers at Kampala got into the air, the landing zone would become a death trap.

June 23, 1999, 2400 Hours

The motorized reconnaissance elements arrived. Everything was on a shoestring. In this case, a total of eight "Hummers" covered a frontage of 12 kilometers, alternately darting forward from one high point to another. Reaching one, they would stop for a few minutes to look, comparing their positions with maps and satellite photos—the former not always agreeing with the latter—and updating their tactical overlays on their IVIS terminals. Along the way, every telephone line encountered was cut—necessarily some were cut more than once—and villages were bypassed. The ground-recon element was halfway to the objective before the first OH-58D scout/attack chopper appeared overhead. Such was the urgency of the mission.

From the pilot's seat of the lead Kiowa Warrior, CWO-4 Jennifer Grayson looked out across moonlit rolling hills dotted with scraggly cotton and cornfields, a few scrawny cattle, and the thatched rooftops of half-deserted villages. It was a far cry from the amber waves of grain that covered her native county back in Kansas. Twenty years of civil war, and the horrendous mortality rates of "slim disease" (the ironic African term for AIDS), had devastated this beautiful country. It had taken two 4th Squadron mechanics and a crew chief less than seven minutes to roll the OH-58D down the ramp of the C-17, bolt on the mast-mounted sight, unfold the rotor blades, and prepare the machine for takeoff. Grayson had offered to help—she knew every centimeter of the graceful little helicopter intimately—but the ground crew had practiced this intricate drill so well that an extra pair of hands would only get in the way. Besides, it was *their* bird after all.

Tonight's mission was to take out the MiGs and Hind-Ds based at Kampala. When his force stormed Entebbe Airport, Colonel O'Connor didn't want any interference from Amin's air force. On the long flight from Fort Polk, Grayson had studied the satellite photos. They were oblique shots, at about 10-centimeter/4-inch resolution, with good lighting and careful image processing to emphasize the details of the revetments and aircraft shelters. Those Russian birds took some nice snaps, she thought.

June 24, 1999, 0100 Hours

The M8 Buford Armored Guns clattered down the ramps of the C-17s and rattled off into the African night, leaving only a whiff of diesel exhaust behind. When the first production units had been delivered the year before, the Army had named them to honor John Buford, the Union cavalry general who had used the repeating rifles of his few dismounted troopers to delay the advance of a Confederate corps on the first day at Gettysburg. It was an eternal lesson every cavalry trooper knew instinctively: Volume of fire carries more weight than superior numbers.

Uganda - Entebbe Airport Action

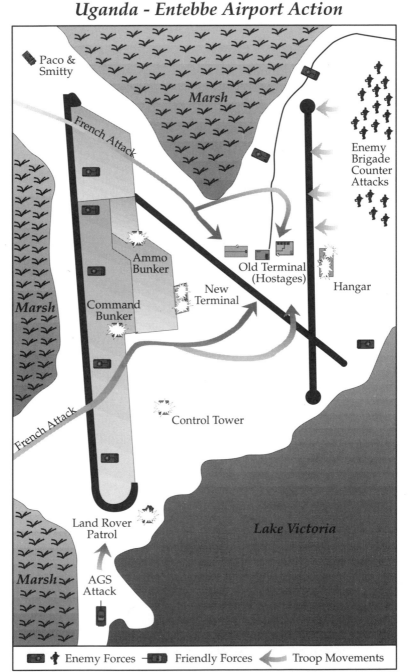

Paco & Smitty

Marsh

French Attack

Enemy Brigade Counter Attacks

Ammo Bunker

New Terminal

Old Terminal (Hostages)

Hangar

Marsh

Command Bunker

French Attack

Control Tower

Lake Victoria

Land Rover Patrol

AGS Attack

Marsh

Enemy Forces — Friendly Forces ← Troop Movements

The assault and rescue by Allied personnel of the hostages at Entebbe Airport, Uganda. While the French special-forces personnel rescue the hostages at the old terminal building, M-8 armored-gun systems and HMMWVs with N-LOS missiles attack the airport security forces. They then stop an attack by an enemy brigade to prevent the evacuation effort.

JACK RYAN ENTERPRISES, LTD., BY LAURA ALPHER

June 24, 1999, 0130 Hours

Sergeant Abu-Bakr Elmahdi cursed in colorful Arabic, Dinka, and English, as the greenish image in his night-vision goggles faded, flickered, and went black. You could walk into any sporting goods store in Europe and pick up a set of the bulky but reliable ex-Soviet goggles for a hundred U.S. dollars. The second-generation image-intensifiers amplified any available light, so that you could make out a man-sized target at a hundred meters by starlight on the blackest night. The Russian engineers had used standard Japanese Ni-Cad camcorder batteries, knowing that the invincible armies of Socialism could procure them anywhere in the world. Unfortunately, a supply clerk back in Khartoum had slyly substituted old, worn-out batteries for the new ones that came with the brigade's equipment. He had sold them in the bazaar for a month's salary apiece. It was outright theft, and it would cost him a hand when it was discovered. But meanwhile, Sergeant Elmahdi and the other sentries along the Entebbe Airport perimeter had only the light of a quarter moon to patrol by.

June 24, 1999, 0200 Hours

Nicole had not been so afraid since the nuns had caught Jean-Jacques hiding in the closet of her dorm room. She never heard from him after that, and only later had learned that his parents had banished him to some military school in the south of France. She smiled thinly as she wondered which of them had suffered more for their sinful behavior. This was worse, though. The Ugandans had herded all of the hostages into the departure lounge of the decrepit old terminal, which still showed bullet scars from the 1976 Israeli rescue mission. There will be no rescue this time, she thought bitterly. The world does not care about a handful of idealistic medical fools who could not turn away from the horror and suffering here. Mercifully, the water taps and the toilets still worked. And twice a day their captors brought baskets of bananas and tubs of cornmeal mush to feed the hostages. But all of their medical supplies and equipment had been looted or destroyed. Some of the male doctors and lab technicians had been badly beaten, though the women had not been molested—yet. Nicole struggled to maintain her composure, to set a good example for the younger girls. Silently and calmly, she recited the Rosary, just as the Sisters had taught her, so many years ago :

Holy Mary, Mother of God,
pray for us sinners,
now, and at the hour of our death.
Amen.

June 24, 1999, 0225 Hours

Smitty eased the HMMWV up just below the crest of a low hill. Paco checked the GPS receiver and confirmed they were in the right place. Off to the south, the vast expanse of Lake Victoria reflected the moonlight, but the town of Entebbe and the airport were blacked out. The airport's diesel generators had been shut down to conserve fuel and deny IR-homing missiles an easy target. Paco thumbed a button and the HMMWV's multi-sensor unit extended on its articulated arm to peer over the top of the hill. He slowly panned the thermal viewer across the airport perimeter. "Bunker . . . three tanks . . . two APCs . . . bunker . . . some kind of SAM launcher . . . six trucks . . . another bunker," he said, carefully enumerating the possible targets, zooming up the magnification to confirm a few doubtful items. Paco looked over at the glowing amber symbols on the IVIS display. Captain Martin, in his command HMMWV two miles back, had sketched out a rough diagram of the airport and designated targets for each unit of the heavy weapon platoon. The 120mm mortars would take out the trucks and lay down smoke. Martin's own HMMWV would take out the airport tower with two N-LOS rounds, to make certain that the commanding four-story building was thoroughly demolished. Smitty and Paco would fire their first N-LOS missile at the second bunker. At the bottom of the screen the captain had written "TOT 2330 Zulu" and underlined it with three bold strokes. All the platoon's weapons were to strike their targets simultaneously at 2330 hours (11:30 PM Greenwich Mean Time, 2:30 AM local time). According to the digital clock on the HMMWV's dashboard, that would be in five minutes. At this range the missile's time of flight would be—Smitty punched a few buttons on the armament control panel—thirty six seconds.

June 24, 1999, 0230 Hours

Ekwanza and Hubutse had drawn patrol duty tonight. The Sudanese lieutenant had screamed at them in broken Swahili, telling them that the countryside was full of American and French spies, and if they let one slip through, they would both die horribly. With its lights off, the Land Rover was to slowly circle the outer perimeter road. Ekwanza had an RPG-7, and Hubutse carried an AKM assault rifle. There was a box of grenades, some signal flares, and a light machine gun in the back of the Rover. They were to be especially vigilant along the lake shore, where the Americans would undoubtedly try to infiltrate Navy SEALs. The lieutenant had been on the cleanup detail a year ago, after the SEALs had paid a nocturnal visit to Port Sudan. He still had nightmares about dark forms rising out of the water...

The gunner of the M8 Buford saw the Rover come around the corner a second or two before Ekwanza and Hubutse saw the AGS looming out of the darkness. Ekwanza was trying to aim his RPG when a 105mm high-explosive (HEAT) round passed through the dead center of the grille and struck the engine block. Land Rovers have a legendary reputation for toughness and reliability in East Africa, but were never meant for that kind of abuse.

"The quarterback is toast!" said the gunner over the intercom exultantly.

"Calm down! Next time use the machine gun on soft targets," the tank commander said. "We've only got twenty rounds of that stuff left now."

The explosion alerted the entire Entebbe garrison. Colonel Alakbar jerked out of bed and knew in an instant what was happening. "The Americans are attacking! Kill all the prisoners at once!" he screamed. The executive officer of the brigade on duty at the command bunker that night was alert and efficient. He was on the field telephone immediately to relay the colonel's order to the company of the 6th Islamic Legion guarding the prisoners. But before he could get the words out, Smitty and Paco's first N-LOS missile came through the sandbagged door of the bunker and detonated against the back wall. The bunker, the colonel, the brigade exec, three sleepy corporals, the field telephone, and the footlocker containing next week's payroll all ceased to exist within a few milliseconds.

June 24, 1999, 0231 Hours

Paco's steady hand guided the second N-LOS missile to a direct hit on a battalion ammo bunker. The secondary explosion was heard in Kampala, twenty-two miles away. "He shoots, he scores!" said Smitty gleefully.

Nicole and all the other hostages woke up when the ammunition bunker exploded 300 yards away.

Though their 150 Sudanese guards had grown up in a traditional warrior society, where killing was part of being a man, and had been brutalized by years of guerrilla warfare in the southern Sudan, where atrocities against civilians had become a routine part of a day's work, they were not evil men. Without orders, they felt uneasy about killing the infidel white doctors. Many of them had an uncle, a cousin, a son, or a grandfather whose life had been saved by people like these. Still, they knew they would probably be ordered to kill the prisoners eventually. If so, it was Allah's will. But without orders, they hesitated. The delay would be fatal for them.

That short hesitation was all the French paras and commandos needed. The heavy weapons of the Eagle Troop had cleared a path for them right up to the old terminal. Their thin-skinned armored cars and personnel carriers converged on the building at high speed from several directions, in a shower of smoke canisters to confuse the defenders and prevent them from getting clear shots. Every man had the new-generation thermal viewer goggles, to see through the smoke and darkness. The hostage-rescue team had brought bullhorns, and over the din of battle they told the hostages to stay down and not move, in French, English, and Arabic. With automatic weapons at point-blank range, it was over quickly. Only a few of the hostages were grazed by stray rounds. No prisoners were taken, and the paras methodically inspected and videotaped with a small camcorder every Sudanese body, putting one round through the head of any that still lived. This done, they turned their attention back to the hostages.

June 24, 1999, 0245 Hours

The Ugandan infantry brigade in the town of Entebbe included a mixed rabble of urban street gangs, guerrilla fighters from the northern tribes, and fanatical Libyan and Sudanese "volunteers." As stragglers from the rout at the airport staggered into his forward outposts, the colonel in charge organized a hasty counterattack to retake the terminal complex. His battery of 122mm field guns knocked out two HMMWVs and an AGS, before the guns were silenced by counter-battery fire from American 120mm mortars and 2.75"/70mm rockets fired by an OH-58D flying top cover over the evacuation. The infantry, charging on foot across open ground, were mown down by interlocking fields of fire from the .50-caliber machine guns of the HMMWVs.

June 24, 1999, 0300 Hours

Now came the hard part. The extraction. Most of the 2nd Cavalry HMMWVs could each carry four extra passengers. The handful of VAB 6x6 armored carriers the French had brought could carry about twelve. Calmly, efficiently, and with a gentleness that was surprising in such a tough-looking warrior, Lieutenant Colonel O'Connor supervised the evacuation of the distraught hostages, making sure each vehicle was properly loaded and that they did not bunch up to create a traffic jam in the rubble-strewn parking lot. As soon as the last enemy snipers had been eliminated from the airport and its surrounding area, he called in his fifteen UH-60L Blackhawk helicopters, which had been waiting back at the landing zone. They evacuated the wounded first, then started shuttling a dozen freed hostages each on the short hop back to the transports.

June 24, 1999, 0330 Hours

A sixteen-man detachment of the 2nd Cav's medical troop had rigged a tiny but well-equipped field hospital in the cavernous interior of one C-17. The freed hostages were haggard, malnourished, and dazed by their sudden rescue, but they were also dedicated professionals. As soon as they arrived, many of the doctors and nurses begged to be allowed to scrub up and help attend to the wounded. The caregivers soon outnumbered the casualties.

Flying high above Lake Victoria, two EH-60 "Quick Fix" Electronic Warfare helicopters from the 4th Air Cavalry Squadron Headquarters and Heaquarters Troop could detect and monitor most of the radio traffic in Uganda that night. The transmitter of Radio Entebbe had been knocked out by a single N-LOS missile, but the enemy brigade in the town still had at least three shortwave sets that had not yet attempted to broadcast a warning to Amin's forces in Kampala. When they tried, they would be located within seconds, jammed immediately, and taken out by precise missile fire soon after. A Swahili instructor hastily assigned from the U.S. Army Language School in Monterey, California, sat at one of the consoles in the crowded cabin, impersonating the disc jockey of Radio Entebbe's late-night program of East African popular music. No one listening would think that anything was amiss in Entebbe. In fact, he mused to himself, his quick run over to the Tower Records in town for a selection of CDs before he had boarded the transport meant that he had the best music collection in this part of Africa. He hoped the locals would appreciate it.

June 24, 1999, 0400 Hours

The pullout had been planned as carefully as the break-in. As soon as the hostages were rescued, the French would race back to the landing zone, emplane, and depart for Djibouti. The light cavalry troops would form a rear guard, falling back from one low ridge to another, while the 4th Air Cavalry Squadron's helicopters covered the withdrawal. Everyone and everything had to be out of Ugandan airspace before the sun rose over Lake Victoria. Damaged vehicles would be blown up in place; there was simply no time to recover them. But, like the Legionnaire code of old, *everyone* would go home.

June 24, 1999, 0410 Hours

When he was awakened by the unforgettable sound of MiG-29s exploding, Halim rolled off his cot, pulled on his coveralls, and ran outside to the helicopter. They had practiced this many times. The lazy Africans might be caught napping, but the Libyan volunteers of the Islamic Jihad Air Unit were determined not to die uselessly on the ground. His pilot, Omar, was already in the pilot seat, starting the engines. As the big five-bladed rotor spun up, Halim jumped into the gunner's seat, strapped himself in, put on the night-vision goggles, and armed the cannon and missile launchers. The big Mi-24 pulled away from the blazing and exploding inferno of Jinja airfield, evaded the fire of the circling AC-130U gunships, and sped toward Entebbe. It would arrive too late.

June 24, 1999, 0415 Hours

Field Marshal and President for Life Al Hajj Idi Amin Dada was wakened by a trembling orderly. The old man had grown increasingly cranky in his seventies, and his six-foot-four-inch frame was still powerful enough to inflict serious injury on a careless underling. "Excellency, we have word that the French and Americans have attacked Entebbe!" the orderly shouted.

"Bring the Mercedes around, and summon General Bashir, we must counterattack immediately!" he roared with the booming parade-ground assurance that had once made him the best regimental sergeant major in the King's African Rifles.

Within the hour, the Presidential Guard Brigade was rolling down the Kampala-Entebbe highway, with a battalion of Libyan T72-M tanks in the lead. General Bashir, the Sudanese chief of staff, who was the actual commander, rode an MTLB command track just behind the tanks. The armored presidential limousine, with an escort of spit-and-polish motorcycle riders, brought up the rear of the long column.

June 24, 1999, 0445 Hours

Colonel O'Connor expected that the enemy would dispatch a relief column from Kampala before morning, despite the radio blackout and deception plan. He had positioned a platoon of Bufords in a good ambush location, with a few helicopters on call to support them with Hellfire missiles. The first salvo of missiles took out the lead tanks, to block the road, and also took out General Bashir's command track, which deprived the brigade of effective leadership. Most of the tanks were knocked out before they could return fire, but the Ugandan veterans of the 2nd Mechanized Battalion, following a few hundred yards behind, piled out of their Chinese APCs and dispersed into the fields along the roadside to bring their hand-held anti-tank weapons to bear on the flank of the ambush force. They knocked out one Buford, but were routed by coaxial machine-gun fire and some .50-cal bursts from the OH-58Ds. Some soldiers kept running until they reached the Sudanese border. A few diehards were cut down as they made a last stand around the presidential limo. O'Connor landed on the road in his HQ Blackhawk, which had finished evacuating the medical team from Entebbe. He yanked open the door of the limo, and was startled for a moment as recognition took hold in his mind. Then, the professionalism of almost two decades in the Army took hold as he grabbed Amin and dragged him out of the car. Almost as if he did this every day of his life, he said, "You are under arrest in the name of the United Nations for crimes against humanity. You have the right to remain silent. . . ."

The victors and their captive boarded the Blackhawk for the long trip back to Dover, Delaware, and eventually, a jail cell in the new UN maximum-security prison complex outside Geneva.

June 24, 1999, 0500 Hours

General du Brigade Jean-Jacques Beaufre personally supervised the loading of the transports that would carry the rescued medical team to the safety of a French air base in Djibouti. He wanted to make sure that his guests were comfortable; and though his rough paras were generally rude to civilians, tonight he would not tolerate that. Then he noticed one of the doctors who was changing the dressing on the head wound of one of the American tankers. The years had etched a few lines in her face and there were strands of gray in her dark hair, but the eyes were unforgettable. She looked up and noticed his insignia of rank. "Jean-Jacques, I see you did well in military school," Nicole said as she smiled.

Tomorrow's Troopers

Why cavalry? In every age the answer has been and will be the same: Commanders need mobile warriors who can scout, screen, engage, and pursue their foes. Whether they ride horses, motor vehicles, flying machines, or devices we cannot yet imagine, as long as there is conflict, there will be a need for cavalry.

Who are the cavalry? Men and women who are drawn to the profession of arms, and who seek out membership in a small, proud, cohesive community of soldiers. The U.S. Army's cavalry is a community that draws strength from tradition, but seems to welcome the best people, innovative ideas, and new technologies from every other combat branch—Infantry, Armor, Aviation, or Artillery. Everyone who has ever seen a classic John Ford Western movie knows who the cavalry are. They are the ones who hold the line on the lawless frontier. They are the soldiers who come to the rescue. Even though some threats have disappeared in the last few years, we still live in a world where there is no shortage of lawless frontiers and people to be rescued.

Professional soldiers and self-appointed experts are always debating what kind of Army we should build. They're always asking questions like: What is the right mix of "light" forces for rapid deployment to low- intensity conflict vs. "heavy" forces like General Franks' VII Corps, which smashed Saddam's Republican Guard? Such questions have been debated for years, and they will continue to preoccupy those who search for (or dream of finding) solutions to the problems of force structure and balance.

But there is a new factor to add to the debate, and it may be decisive in determining what kind of Army will be built in the next ten to twenty years. That force is high technology.

Technology has always been a factor in deciding how to equip and organize armed forces. Ever since the first man chose to pick up a rock or stick to gain advantage over other men, there has been a race to find better rocks and sticks. And when these couldn't be found, men designed and built new and improved rocks and sticks.

Once upon a time, it took ten years or more to get a weapons system from the drawing board to the battlefield. No longer. Today, the bewildering pace of technological change makes it hard to decide what kind of rocks and sticks to build.

What has suddenly changed is the availability of programmable digital systems. The revolutionary aspect of these digital weapons systems is that much of their performance is based upon lines of programming code. They have a built-in growth potential. Rewrite a few software modules, and change out some hardware packages on the data bus, and your old system becomes a new weapon with vastly improved capabilities. One only need look at the M1A2 Abrams and the AH-64D Longbow Apache to see the truth of this. But to make all this technology work, you need soldiers with the mental fitness and agility to adapt to constant change. You need cavalry troopers.

It is said that an Army always prepares for the last war. Thanks to Saddam Hussein, today our Army has plenty of experience, equipment, and training for desert fighting. So, given the reality of Murphy's Law, we can probably expect our next war to be in wooded mountains or urban jungles. Again: You need cavalry troopers.

The personnel of the 2nd ACR-L and the 3rd ACR are going to be the vanguard of a new generation of cavalry troopers. They will be armed with an array of high-technology equipment that offers tactical options their predecessors in the Gulf War could only have read about in science fiction novels. Individual troopers will become a part of large-scale computer networks like the IVIS system. Stealth systems, like tanks and fighting vehicles made of composite plastics, will make their presence known in ways that we cannot imagine. The commander's task will be to sort through a flood of data to find the nuggets of tactical opportunity that are presented. That is the challenge that the cavalry leaders of the 21st century are going to face.

Some of the brightest people in America's Army have been rethinking the "roles and missions" that this world will require of them over the next few decades. It will be a world where religious extremism and ethnic hatred have increasing access to weapons of mass destruction. It will be a world challenged by threats of ecological terrorism, bioengineered plagues, and widespread social and economic breakdown. And increasingly, all of America's armed forces will be called on to take part in operations other than war. Some of these include:

- Disaster relief and reconstruction (floods, earthquakes, fires, famines, etc.)
- Counter-narcotics operations (intervention in drug-producing regions, and support of law enforcement on our own borders)
- Peacekeeping and peace enforcement (under United Nations authority, or in our own cities)

All of these operations require land and air mobility for rapid deployment, as well as massive firepower to deter armed thugs and maintain or restore order. If use of armed force becomes necessary, the ability to make quick decisions, improvise, and solve problems at the level of squad and platoon leadership will be vital. These are defining characteristics of the cavalry, and show why preserving it as an American military institution is so vital. As we have seen in Somalia, even a humanitarian relief mission can turn suddenly into an armed conflict. As we have seen in Bosnia, the lack of peace-keepers with overwhelming firepower and advanced technology can leave the most well-meaning intervention powerless to prevent genocide.

A recurrent cycle in American history has been victory in war followed by such heedless and rapid disarmament that the next war has caught us unprepared. In the late 1940s, the army that helped crush Hitler's Wehrmacht and was prepared to invade Japan against last-ditch *kamikaze* resistance was gutted in the mistaken belief that it would not ever again be needed. Only five years later, American troops (the ill-fated Task Force Smith[1]) were routed by the forces of a fourth-rate power—North Korea. Fortunately, we have always kept a few cavalry regiments on hand. Having met some of today's cavalry troopers and seen what they can do, you may agree that our national investment in tomorrow's troopers is a wise investment indeed. Armed forces are expensive. Superbly trained and equipped armed forces like the 3rd ACR are *very* expensive. But the only thing more costly to a nation is not having them when you need them.

Glossary

AAR After-Action Review. A critical evaluation of participant performance in an exercise or simulation.

ACR Armored Cavalry Regiment.

AH-1 Huey Cobra, U.S. attack helicopter.

AH-56 Cheyenne, cancelled U.S. attack helicopter. The predecessor of the Apache.

AH-64 Apache, U.S. attack helicopter.

AHRS Attitude-Heading Reference System. The primary navigational system of the Apache attack helicopter.

APC Armored Personnel Carrier.

APFSDS Armor-Piercing, Fin-Stabilized, Discarding-Sabot round. See **Long-Rod Penetrator**.

AOBC Armor Officer Basic Course. Located at Fort Knox, Kentucky.

AT-# NATO designation system usually used to identify Soviet ATGMs. Unfortunately, some NATO weapons are also starting to use this designation, which leads to great confusion.

ATACMS Army Tactical Missile System.

ATGM Anti-Tank Guided Missile.

BCS Battery Control System.

BDU Battle Dress Uniform. The new term for camouflage uniforms.

Bill A Swedish heavyweight anti-tank missile that uses an angled shaped-charge warhead to attack a tank's top. The name is derived from a medieval hooked-pole weapon, not the nickname for "William."

Blazer The name given to Israeli first-generation explosive reactive armor.

BMP-3 A Russian Infantry Fighting Vehicle.

C^3I	Command, Control, Communications, and Intelligence. Pronounced "See-Three-Eye."
CENTCOM	Central Command. The unified command that was in charge of Operations Desert Shield/Storm/Saber.
Ceramic Laminates	Tank armor that incorporates a ceramic and composite layer or layers to improve the armor's resistance to shaped-charge rounds.
Challenger I	A 1980s-era British main battle tank. First British tank to use the special Chobham armor.
Challenger II	The latest variant of the British Challenger series. Improvements include new armor, second-generation Chobham, a 120mm gun, and more sophisticated vehicle electronics.
Chobham Armor	A revolutionary type of armor, developed by the British Army research facility at Chobham, England, that uses standard RHA steel sections and a special ceramic composite layer in between. Chobham armor is very resistant to penetration by shaped-charge jets.
CID	Commander's Integrated Display.
CINC	Commander In Chief.
CITV	Commander's Independent Thermal Viewer.
Combination Armor	The Soviet/Russian term for armor that uses layers of rolled homogeneous steel and ceramic laminates.
Comanche	A new U.S. scout/attack helicopter in development, the RAH-66.
Copperhead	Designed for use against tanks and point targets, the Copperhead is the artillery equivalent of a laser-guided bomb.
CP	Command Post.
CS	Tear Gas.
DID	The Driver's Integrated Display on the M1A2.
Dragon	An optically tracked, man-portable anti-tank missile.
DU	Depleted Uranium. Natural uranium that has had most of the uranium-235 isotope removed. Almost pure uranium-238.
ECM	Electronic CounterMeasures.
EFP	Explosively Formed Plate. The EFP or flying plate uses, like

the shaped charge, a high explosive to deform a metallic liner into a projectile. Shaped-charge warheads use a conical-shaped liner, whereas the EFP uses a shallow hemispherical dish. When the explosive detonates, it deforms the dish into a solid slug. The slug is propelled at speeds up to Mach 5.

ERA Explosive Reactive Armor. An add-on armor system that uses high explosive sandwiched in between steel plates. The explosive detonates upon impact from a shaped-charge jet or long-rod penetrator. First-generation ERA was designed to reduce the effectiveness of a shaped- charge jet by breaking it up. Second-generation ERA provides improved performance against shaped-charge jets, but can also break long-rod pene-trators. Once used, the ERA tile has to be replaced.

FAADS Forward-Area Air-Defense System. A Bradley variant that replaces the TOW launcher with four Stinger SAMs.

FASCAM Field Artillery Containerized Anti-Tank Mine.

FCEU Fire-Control Electronics Unit.

FLIR Forward-Looking InfraRed.

Fratricide Casualties caused by "friendly fire" or friendly units firing on each other.

FV-432 Warrior A British Army infantry fighting vehicle.

GCDP The Gunner's Control and Display Panel on the M1A2.

GPS Global Positioning System. A series of satellites that can accurately provide a unit's location (within several hundred feet of its true location).

HE High Explosive.

HEAT High-Explosive Anti-Tank. See **Shaped-Charge Warhead**.

Hellfire A U.S. long-range, laser-guided anti-tank missile that is the main armament of many Western attack helicopters (AGM-114).

Hellfire II The next-generation Hellfire missile which uses a millimeter-wavelength radar seeker. The Hellfire II is a "brilliant" weapon, because it can discriminate between the different types of targets with its advanced seeker.

HETS Heavy-Equipment Transporter System. A big tractor-trailer that can haul large vehicles including M1 series tanks.

HEU Hull Electronics Unit.

HE-VT	High-Explosive, Variable-Time fuse.
HMMWV	High-Mobility Multipurpose Wheeled Vehicle. The "Jeep" of the 1980s/90s.
HUD	Heads-Up Display.
IFV	Infantry Fighting Vehicle.
IVIS	Inter-Vehicular Information System. The new computer network that is installed in M1A2 MBTs.
Javelin	The first "fire and forget," shoulder-launched anti-tank guided missile to enter service anywhere in the world. With Javelin, there are no wires since the missile has an advanced imaging infrared seeker which is locked on to the target before firing.
J-STARS	A reconnaissance aircraft, based on the Boeing 707 airframe (called an E-8), carrying a synthetic aperture radar to view ground targets from ranges of hundreds of miles. Able to transmit the information to divisional and corps ground stations.
JTF	Joint Task Force.
Kiowa Warrior	OH-58D scout attack helicopter.
Leopard I	A 1970s-era German main battle tank.
Leopard II	A 1980s-era German main battle tank. Has the same offensive capability of the M1 series, but has a poorer armor package that doesn't use Chobham armor.
Longbow	A millimeter-wave radar, designed to see ground and air targets in any type of weather, day or night. To be fitted on D model of the AH-64 Apache attack helicopter.
Long-Rod Penetrator	The modern kinetic-energy round which resembles a 1.5-to-2-foot (.46-to-.61-meter)-long, 10-lb (4.54-kg) metal dart. These darts can achieve speeds up to Mach 4 when shot from a large-caliber tank gun. The long-rod penetrator defeats a tank's armor by application of brute force to a very small area of the armor's surface.
LZ	Landing Zone. An area designated for landing airborne (parachute) or airmobile (helicopter) troops.
M1 Abrams	A 1980s-era U.S. main battle tank. Follow-on to the M60. First U.S. tank to use the special Chobham armor.
M1A1	Improved M1 tank with a better armor package and the incorporation of the Rheinmetall 120mm gun.

M1A1 HA	The Heavy Armor variant of the M1 series. This tank includes depleted uranium in the armor package to help improve the armor's effectiveness against long-rod kinetic-energy penetrators.
M1A2	The latest version of the M1 series. Includes all the improvements in the M1A1 HA plus new sensors and a more sophisticated internal computer system.
M2/M3 Bradley	A 1980s-era U.S. Infantry Fighting Vehicle.
M4 Sherman	A 1940s-era U.S. tank that saw considerable action in World War II. While it was not the technological equal of most German designs, it was a decent tank that won battles because the U.S. could make a lot of them in comparison to the more complex German tanks.
M26 Pershing	A 1940s-era U.S. tank. Follow-on to the M4 Sherman.
M48 Patton	A 1950s-era U.S. main battle tank. First post-World War II U.S. heavy tank.
M60	A 1960s-era U.S. main battle tank. Follow-on to the M48.
M106	A mobile 81mm mortar carried on an M113 chassis.
M109A6	Paladin 155mm Self-Propelled Howitzer.
M113	A 1960s-era U.S. armored personnel carrier.
M125	A mobile 106mm mortar carried on an M113 chassis.
M270	The combined MLRS carrier/launcher vehicle.
M577	A mobile command post based on an M113 chassis with a raised roof and sides, and additional generators to power the numerous radios that are stored in racks inside the rear compartment.
M901	M113 chassis with an erectable two-round launcher and optical sight for TOW anti-tank missiles.
M993	The chassis that carries the MLRS rocket packs.
Maverick	An air-launched U.S. laser or infrared-guided, air-to-surface missile that can be used in an anti-tank role (AGM-65).
MBT	Main Battle Tank.
MREs	Meals Ready to Eat. The C-rations of the 1980/90s. Sometimes humorously referred to as Meals Rejected by the Enemy/Everyone/ Ethiopia.
Merkava	A 1980s-era Israeli main battle tank.

MFDs	Multi-Function Displays.
MILES	Multiple Integrated Laser Exercise System. MILES provides a "no-shoot" method to practice ground combat (and some air combat), while scoring and recording results. MILES uses eye-safe lasers to simulate the firing of weapons and laser detectors to record "hits."
MLRS	Multiple-Launch Rocket System. A stretched M2 Bradley hull with two six-cell rocket packs.
MMS	Mast-Mounted Sight. An optical sensor package on OH-58D Kiowa Warrior helicopters that includes a stabilized FLIR, a daylight television camera, as well as a laser range finder and designator.
MOS	Military Occupational Specialty. A designation system that identifies a soldier's specialized training.
NATO	North Atlantic Treaty Organization.
NBC	Nuclear, Biological, Chemical.
NCO	Non-Commissioned Officer.
NOTAR	No-TAil Rotor. A helicopter design that uses a forced air system to counteract the torque placed on the helicopter from the main rotor.
NTC	National Training Center. Located at Fort Irwin, California.
NVG	Night-Vision Goggles.
OPFOR	Opposing Forces. The "bad guys" in a training exercise.
Pave Low	MH-53J Air Force special-operations helicopters for communications, navigational, and rescue support.
PNVS	Pilot Night-Vision Sensor.
POW	Prisoner Of War.
RAM	Radar-Absorbent Material. One of the components of "stealth technology." RAM absorbs and dissipates radar waves, rather than reflecting them back toward a detector.
RCS	Radar Cross-Section. A measurement of how much radar energy an object reflects in a particular direction.
RDF	Radio Direction-Finding.
REFORGER	REturn of FORces to GERmany. The annual NATO exercise to test rapid Allied reinforcement of forces in Europe

in anticipation of an all-out attack of Soviet forces into West Germany.

Republican Guard The elite Iraqi Army force, originally Saddam Hussein's personal guard, later grew to over 100,000 troops. These units were the best armed of the Iraqi Army and were considered to be the most dangerous by Coalition forces.

RHA Rolled Homogeneous Armor. A family of high-quality steel alloys, rolled to provide a uniform thickness as well as the best combination of strength and penetration resistance. The material is uniformly hard throughout and it is the standard by which other armor types are evaluated.

RIU Radio Interface Unit.

ROE Rules Of Engagement. Specific listed conditions under which U.S. military forces may fire on a hostile unit.

RPG Rocket-Propelled Grenade. A series of Soviet/Russian light-weight, disposable anti-tank weapons.

RPV Remotely Piloted Vehicles. These are small, unmanned, remotely controlled reconnaissance drones.

RWR Radar Warning Receiver.

SA-# NATO designation system usually used to identify Soviet SAMs.

SADARM Sense-And-Destroy Armor. A special munition which can sense the presence of an armored vehicle or artillery piece and then attack it with EFP warheads.

SAM Surface-to-Air Missile.

SCUD A NATO nickname for an early-generation, unguided tactical ballistic missile.

Shaped-Charge Warhead A shaped-charge warhead is a high-explosive charge packed around a cone-shaped metallic liner. When detonated, the explosive charge causes the metal liner to collapse rapidly towards the center of the round. The metal liner is heated and compressed by the energy of the explosion to form a jet with velocities as high as Mach 25. The metal remains solid but behaves like a fluid because of the tremendous pressure. Virtually all ATGMs have a shaped-charge warhead.

SINCGARS SIngle-Channel Ground and Airborne Radio System. A family of frequency-hopping FM radios that can use any of 2,320

	different frequencies between 30 and 87.975 MHz in the VHF band.
Spall	Parts or fragments of the tank's armor that are broken off and blasted into the interior by a hit from an anti-tank or artillery round. Spalling can occur even when the armor is not fully penetrated.
SPH	Self-Propelled Howitzer.
SSM	Surface-to-Surface Missile.
Stinger	A man-portable, infrared-guided surface-to-air missile. Stingers can also be fired from land vehicles or helicopters with modified launchers.
T-64	A 1970s-era Soviet main battle tank that incorporated the first combination/ceramic laminate armor. Follow-on to the more conventional T-62.
T-72	A 1970s-era Soviet main battle tank that was designed to be a simpler version of the T-64 so that it could be mass-produced. Widely exported to the former Warsaw Pact and Soviet client states.
T-80	A 1980s-era Soviet/Russian main battle tank. This is the best tank in the Russian inventory today and is not far below the M1A1 in performance.
TACAN	Tactical Navigational System.
TACFIRE	Tactical Fire-Direction Computer. An Army artillery fire-control coordination system being replaced by the Advanced Field Artillery Tactical Data System.
TACOM	Tank and Automotive Command. Located in Warren, Michigan.
TADS	Target-Acquisition Designation Sight.
TCP	Tactical Command Post.
TGW	Terminally Guided Warhead. A new-generation smart munition which is expected to be ready for low-rate production by 1996. The TGW is a large anti-armor munition which uses a "smart" millimeter-wave radar seeker to search out tanks and other priority targets.
TIS	Thermal-Imaging Sight.
TOC	Tactical Operations Center.

TOW	Tube-launched, Optically tracked, Wire-guided missile. A U.S. heavyweight anti-tank missile which can be fired from either land or airborne platforms. The latest version, TOW-2B, is a top-attack weapon with two angled warheads.
TRADOC	U.S. Army's Training and Doctrine Command, headquartered at Fort Monroe, Virginia.
UH-1	Huey, U.S. utility/transport helicopter.
UH-60A	Blackhawk, U.S. utility/transport helicopter.
WP	White Phosphorous. A type of incendiary artillery shell.
XM4	New U.S. Mobile Command Post tracked vehicle. Replacement for the old M577 mobile command post.
XM5	New U.S. Electronic Fighting Vehicle. This vehicle is designed to scan for enemy radios or radars, determine the bearing and possibly the location of the transmitter, and jam them if necessary.
XM8	Armored Gun System. A new lightweight tank, armed with a 105mm gun, specifically designed for quick-reaction, light forces.
ZSU-23-4	Soviet Shilka mobile air-defense gun. Four radar-controlled 23mm cannons on an armored chassis designed to provide close air defense to Soviet troop formations.
ZU-23	Soviet twin 23mm automatic cannon on ground mount with optical sight. Belt-fed, it can fire over 800 rounds per minute from each barrel out to an effective range of 2,500 meters.

End Notes

Introduction

[1] Purists will refer to the 3rd ACR as the 3rd *U.S.* Cavalry, probably harking back to the Civil War, when states raised regiments identified with their home jurisdictions (1st Michigan, etc.). Thus "U.S. Cavalry" denotes the Regular Army.

[2] Since deactivated.

There and Back Again: An Interview with General Fred Franks

[1] Some people thought that was a bad idea, but a small group of visionaries kept the idea alive long enough to bring it to combat in 1965.

[2] The Division-86 study, initiated by General Donn Starry in 1978 and implemented during the 1980s, reorganized the Army's "heavy" armored and mechanized infantry divisions. Armored divisions would have six tank and four infantry battalions, while mechanized divisions were to have five of each type. It strengthened battalions to four companies, instead of three. It added an attack helicopter brigade to each division, increased the size of the howitzer battery, and specified other changes to increase combat power.

[3] The Soviet 8th Guards Combined Arms Army consisted of three motor rifle divisions, one tank division, and supporting units of artillery, engineers, and attack helicopters. It was one of nine Soviet armies in forward offensive positions along the former border between East and West Germany.

[4] REFORGER (REturn of FORces to GERmany) was a major NATO exercise in Europe. This annual operation, a hallmark of the Cold War years, involved mass movements of personnel and equipment to Germany and other NATO nations. It was designed to help all NATO forces simulate how they would have to rapidly build up and reinforce their forces as a prelude to war with the Soviet Union and the Warsaw Pact. While some of the reinforcing units would bring their heavy equipment via ship, most of the units would obtain their equipment from pre-positioned stocks in Europe.

[5] Plans and Staff Officer.

[6] A division is commanded by a major general (two stars).

[7] A corps is typically commanded by a lieutenant general (three stars). VII Corps was located mainly in the West German state of Bavaria.

[8] Saddam's Republican Guard consisted of eight divisions totaling about 100,000 men. These received the best available recruits, supplies, training, and equipment and were considered the most reliable element of the Iraqi Army. During the war, four Republican Guard divisions were virtually destroyed. Saddam kept the strongest and most loyal division in reserve around Baghdad, and it never saw action.

[9] GPS uses satellites and portable receivers. These tell you where you are with extreme accuracy.

[10] "Force-oriented" means that VII Corps' goal was the destruction of enemy units themselves, not capture of a geographic objective.

[11] Army tactical radios typically have ranges of 8 to 35 km (5 to 22 miles). In flat terrain, relay stations or "repeaters" are usually positioned every 16 to 24 km (10 to 15 miles).

[12] As in football.

[13] Medina Ridge was the name American troops gave to a low rise, about 7 miles long, in the Iraqi desert north of Kuwait, where the First Armored Division's Second ("Iron") Brigade destroyed a brigade of the Iraqi Republican Guard's Medina Division (sixty T-72 tanks and "dozens" of personnel carriers) in forty minutes on the afternoon of 27 February 1991.

[14] FM 100-5 (June 1993) defines a *hasty attack* as one launched "with the forces at hand and with minimum preparation to destroy the enemy before he is able either to concentrate or establish a defense." Such an attack "enhances agility at the risk of losing synchronization." Hasty attack is contrasted with *deliberate attack*, which takes more time to prepare.

[15] FM 100-5 defines *pursuit* as "an offensive operation against a retreating enemy force." *Exploitation* is the follow-up to a successful attack. "Exploitations and pursuits test the audacity of soldiers and leaders alike. Both of these operations risk disorganizing the attacker nearly as much as the defender."

[16] Double envelopment is a simultaneous maneuver against both flanks of an enemy position. It was first used by Hannibal to crush the Roman army at Cannae in 216 BC, and has traditionally been regarded as the ultimate expression of generalship.

Honing the Razor's Edge

[1] A linear sheaf is a particular type of artillery impact pattern.

A Cavalry Officer's Life

[1] Captain McMaster's troop used color code names. Red for 1st Platoon, White for 2nd Platoon, Blue for 3rd Platoon, Green for 4th Platoon, and Black for the Troop Command Element. Captain McMaster's tank call sign was Black-66.

Roles and Missions: The ACR in the Real World

[1] An extended-range version of the Soviet SCUD, the Nodong ballistic missile has a range of 1,000 km (620 miles) with a 1,000-kg (2,200-pound) warhead. With an accuracy ("Circular Error Probable") of perhaps 500 meters (1,640 feet), it threatens all of South Korea and many of the cities in Japan, China, and Siberian Russia.

[2] Note that all times are given in terms of Seoul's time zone, which is fourteen hours ahead of Washington, D.C., and nine hours ahead of Greenwich Mean Time (GMT), which the U.S. Army likes to use for reference worldwide and for some obscure reason calls "Zulu" time. When it is 7:00 AM in Washington, it is 9:00 PM in Seoul, and 1200 Zulu.

[3] The U.S. Eighth Army includes all Army forces in Korea and Japan—totaling 25,000 troops in 1997. Eighth Army Headquarters in Korea is commanded by a four-star general who is also nominal Supreme Commander of all United Nations forces on the peninsula. In practice this means he coordinates planning, logistics, intelligence, and operations with the South Korean military command structure.

[4] Pacific Command (PACOM) in Honolulu, Hawaii, is commanded by a four-star Navy admiral and has operated control over virtually all U.S. forces in the Pacific region, including Eighth Army in Korea.

[5] Weighing 40,000 to 45,000 tons at full load, these huge, boxy cargo ships are loaded with military equipment, maintained on long-term lease in

secure harbors near potential trouble spots, and operated by mixed Navy/contractor crews. A typical MPS can carry 522 standard 20-foot vans (350 ammunition drums, and thirty-two refrigerated) plus roll-on/roll-off parking space for 110 general supplies, thirty with fuel up to 1,400 HMMWV-sized vehicles and 1,500,000 gallons (5,764 cubic meters) of bulk fuel that can be off-loaded. The diesel-powered ships can make 17 knots/31.5kph. These should not be confused with the speedier SL-7 fast transports.

6 The 3rd Armored Cavalry Regiment is normally assigned to the III Corps in Texas. Under the Robust Screen contingency plan described here, it would be transferred, with all of its attachments, to the operational control of the lieutenant general (three stars) commanding IX Corps in Korea.

7 Manned by small civilian crews, they proved their worth in 1990 and 1991 during operation Desert Shield. Just seven of these ships carried 11% of *all* the U.S. cargo transported to the Persian Gulf (the other 89% came mostly by slower chartered transport ships, with only the most urgent cargo airlifted at great cost).

8 1st MEF included three infantry brigades, squadrons of attack and transport helicopters, some light armor battalions, and an air wing with squadrons of F-18 Hornets and AV-8B Harriers.

9 Time references are given in terms of Uganda local time, which is three hours ahead of GMT, and eight hours ahead of Washington.

Tomorrow's Troopers

1 Task Force Smith was a detachment of the 1st Battalion of the 21st Infantry Regiment, reinforced with a field artillery battery. It was rushed to Korea shortly after the North Korean Army launched its invasion of the South on June 25, 1950. On July 5, lacking effective anti-tank weapons, the ill-trained and poorly led task force was overrun and destroyed near Osan. In American military history it has become a symbolic object lesson of how *not* to train, equip, and commit troops to battle.

Bibliography

Official Army Publications

Army Modernization Plan. Annex A,-*Close Combat-Heavy.* U.S. Army, 1993.

Army Modernization Plan. Annex B,-*Close Combat-Light.* U.S. Army, 1993.

Army Modernization Plan. Annex C,-*Command, Control & Communications.* U.S. Army, 1993.

Army Modernization Plan. Annex D,-*Engineer And Mine Warfare.* U.S. Army, 1993.

Army Modernization Plan. Annex E,-*Air Defense.* U.S. Army, 1993.

Army Modernization Plan. Annex F,-*Tactical Wheeled Vehicles.* U.S. Army, 1993.

Army Modernization Plan. Annex G,-*Fire Support Systems.* U.S. Army, 1993.

Army Modernization Plan. Annex H,-*Theater Missile Defense.* U.S. Army, 1993.

Army Modernization Plan. Annex I,-*Intelligence/Electronic Warfare.* U.S. Army, 1993.

Army Modernization Plan. Annex J,-*Logistics.* U.S. Army, 1993.

Army Modernization Plan. Annex K,-*Soldier.* U.S. Army, 1993.

Army Modernization Plan. Annex L,-*Aviation.* U.S. Army, 1993.

Army Modernization Plan. Annex M,-*Nuclear, Biological, and Chemical.* U.S. Army, 1993.

Army Modernization Plan. Annex N,-*Information Mission Area Infrastructure.* U.S. Army, 1993.

Army Modernization Plan. Annex O,-*Medical.* U.S. Army, 1993.

Army Modernization Plan. Annex P,-*Training.* U.S. Army, 1993.

Army Modernization Plan.-Volume I. U.S. Army, 1993.

Army Modernization Plan.-Volume II. U.S. Army, 1993.

Command and General Staff College, Fort Leavenworth, KS. Student Text 100-3, *Battle Book: Center for Army Tactics.* U.S. Army, 1986.

Command and General Staff College, Fort Leavenworth, KS. Student Text 100-7, *Soviet Army Handbook.* U.S. Army, 1991.

Field Artillery School, Fort Sill, OK. *Advanced Field Artillery Tactical Data System (AFATDS) Operations.* U.S. Army, 1992.

Field Artillery School, Fort Sill, OK. *Tactics, Techniques, and Procedures for the M109A6 (Paladin) Howitzer: Section Platoon Battery and Batallion.* U.S. Army, 1992.

FM 1-114, *Tactics, Techniques, and Procedures for the Regimental Aviation Squadron.* U.S. Army, 1991.

FM 17-47, *Air Cavalry Combat Brigade.* U.S. Army, 1982.

FM 100-5, *Operations,* U.S. Army, HQ Training and Doctrine Command, 1993.

FM 100-23, *Peace Operations.* Draft, U.S. Army, 1993.

Headquarters, Training and Doctrine Command. FM 3-4, *NBC Protection.* U.S. Army, 1984.

Headquarters, Training and Doctrine Command. FM 15-50, *Attack Helicopter Operations.* U.S. Army, 1984.

Headquarters, Training and Doctrine Command. FM 55-50. *Army Water Transport Operations.* U.S. Army, 1985.

Headquarters, Training and Doctrine Command. FM 100-17, *Mobilization, Deployment, Redeployment, Demobilization.* U.S. Army, 1992.

How They Fight. Desert Shield Order of Battle Handbook, U.S. Army, September 1990.

Land Warfare in the 21st Century. U.S. Army, 1993.

Product Manager, M113/M60 Family of Vehicles. *Data Book: September 1992.* U.S. Army, Tank Automotive Command, Warren, MI, 1992.

2d Armored Cavalry 1989-1991, 2nd Armored Cavalry Regiment, 1991.

24th Mechanized Infantry Division. *Operation Desert Storm: Attack Plan– OPLAN 91-3.* U.S. Army, 1992.

24th Mechanized Infantry Division. *24th Mechanized Infantry Division Combat Team: Historical Reference Book.* U.S. Army, 1991.

24th Mechanized Infantry Division. *The Victory Book: A Desert Storm Chronicle.* U.S. Army, 1991.

Books

Adams, James. *Secret Armies*. Atlantic Monthly Press, 1987.

Adan, Avraham (Bren). *On the Banks of the Suez*. Presidio Press, 1980.

Albrecht, Gerhard, and Rhades, Jurgen. *Weyers Flottentaschenbuch– Warships of the World*. Bernard & Graefe Verlag, Bonn, 1992/93.

Allen, Thomas B. *War Games-The Secret World of the Creators, Players, and Policy Makers Rehearsing World War III Today*. McGraw-Hill, 1987.

Antal, John F. *Armor Attacks-The Tank Platoon*. Presidio Press, 1991.

Arrian. *The Campaign of Alexander*. Dorset Press, 1971.

Asher, Jerry, and Hammel, Eric. *Duel for the Golan-The 100-hour Battle That Saved Israel*. William Morrow and Company Inc., 1987.

Atkinson, Rick. *Crusade-The Untold Story of the Persian Gulf War*. Houghton Mifflin Company, 1993.

Avallone, Eugene A., and Baumeister, Theodore III (ed.). *Marks' Standard Handbook for Mechanical Engineers*. 9th ed. McGraw Hill Book Company, 1987.

Baxter, William P. *Soviet AirLand Battle Tactics*. Presidio Press, 1986.

Bishop, Chris, and Donald, David. *Encyclopedia of World Military Power*. The Military Press, 1986.

Blackwell, James. *Thunder in the Desert-The Strategy and Tactics of the Persian Gulf War*. Bantam Books, 1991.

Blair, Arthur H. *At War in the Gulf-A Chronology*. Texas A&M University Press, 1992.

Blair, Clay. *The Forgotten War-America in Korea 1950-1953*. Times Books, 1987.

Bolger, Daniel P. *Dragons at War 2-34th Infantry in the Mojave*. Presidio Press, 1986.

Bradley, John H. *The Second World War: Asia and the Pacific*. Avery Publishing Group, 1989.

Bradnock, Robert. *South Asian Handbook*. Prentice Hall, 1992.

Brady, George S., and Clauser, Henry R. (ed.). *Materials Handbook*. 13th ed. McGraw Hill Book Company, 1991.

Brugioni, Dino A. *Eyeball to Eyeball: The Inside Story of the Cuban Missile Crisis*. Random House, 1990.

Buell, Thomas B.; Franks, Clifton R.; Hixson, John A.; Mets, David R.; Pirnie, Bruce R.; Ransone, James F. Jr.; and Stone, Thomas R.. *The Second*

World War: Europe and the Mediterranean. Avery Publishing Group, 1989.

Chadwick, Frank. *Gulf War Fact Book.* GDW, 1991.

Charlton, James (ed.). *The Military Quotation Book.* St. Martin's Press, 1990.

Chinnery, Philip D. *Life on the Line.* St. Martin's Press, 1988.

Clancy, Tom. *Submarine: A Guided Tour Inside a Nuclear Warship.* Berkley Books, 1993.

Cohen, Elliott, and Gooch, John. *Military Misfortunes: The Anatomy of Failure in War.* Free Press, 1990.

Courtney-Green, P. R. *Ammunition for the Land Battle: Land Warfare.* Brassey's New Battlefield Weapons Systems and Technology Series, Volume 4. Brassey's (UK) Ltd., 1991.

Crowe, William J. Admiral Jr. *The Line of Fire.* Simon & Schuster, 1993.

Darwish, Adel, and Alexander, Gregory. *Unholy Babylon–The Secret History of Saddam's War.* St. Martin's Press, 1991.

Deighton, Len. *Blood, Tears and Folly: An Objective Look at World War II.* Harper Collins, 1993.

Delbruck, Hans. *Warfare in Antiquity.* University of Nebraska Press, 1975.

Department of Defense. *Conduct of the Persian Gulf War.* Government Printing Office, 1992.

Dorr, Robert F. *Desert Shield —The Build-Up: The Complete Story.* Motorbooks International, 1991.

——*Desert Storm Air War.* Motorbooks International, 1991.

——*Desert Storm Ground War.* Motorbooks International, 1991.

——*Desert Storm Sea War.* Motorbooks International, 1991.

Dunnigan, James F., and Bay, Austin. *From Shield to Storm.* William Morrow and Company, Inc., 1992.

Dunnigan, James F., and Macedonia, Raymond M. *Getting It Right: American Military Reforms after Vietnam to the Gulf War and Beyond.* William Morrow and Company, Inc., 1993.

Dupuy, R. Ernest, and Dupuy, Trevor N. *The Harper Encyclopedia of Military History from 3500 BC to the Present.* 4th ed. Harper Collins Publishers, Inc., 1993.

Dupuy, Trevor N. *Numbers, Predictions & War: The Use of History to Evaluate and Predict the Outcome of Armed Conflict.* Hero Books, 1985.

——*The Evolution of Weapons and Warfare*. Bobbs-Merrill, 1980.

——*Understanding War: History and Theory of Combat*. Paragon House, 1987.

——*Understanding Defeat*. Paragon House, 1990.

——*How to Defeat Saddam Hussein*. Warner Books, 1991.

——*Future Wars: The World's Most Dangerous Flashpoints*. Sidwick & Jackson Ltd., 1993.

Dupuy, Trevor N.; Johnson, Curt; and Bongard, David. *Harper Encyclopedia of Military Biography*. Harper Collins, 1992.

Edwards, John E. *Combat Service Support Guide*. 2nd ed. Stackpole Books, 1993.

Ellis, John. *Brute Force–Allied Strategy and Tactics in the Second World War*. Penguin Group, 1990.

Eshel, David. *The U.S. Rapid Deployment Forces*. Arco Publishing, 1985.

Everett-Heath, E. J.; Moss, G. M.; Mowat, A.W.; and Reid, K. E. *Military Helicopters, Land Warfare*. Brassey's New Battlefield Weapons Systems and Technology Series, Volume 6. Brassey's (UK) Ltd., 1990.

Farrar, C. L., and Leeming, D.W. *Military Ballistics–A Basic Manual*.

Brassey's Battlefield Weapons Systems and Technology Series, Volume 10, Brassey's Publishers Ltd., 1983.

Foss, Christopher F. (ed.). *Jane's Armour and Artillery 1979-80*. Jane's Information Group, 1979.

——*Jane's Armoured Personnel Carriers*. Jane's Publishing Company Ltd., 1985.

——*Jane's Main Battle Tanks*. 2nd ed. Jane's Publishing Company Ltd., 1986.

Friedman, Norman. *Desert Victory: The War for Kuwait*. Naval Institute Press, 1991.

——*World Naval Weapons Systems*. Naval Institute Press, 1991.

Gilbar, Stephen. *The Reader's Quotation Book*. Penguin, 1990.

Glover, Thomas J. *Pocket Ref.* Sequoia Publishing, 1992.

GPS: A Guide to the Next Utility. Trimble Navigation, 1993.

Green, Michael. *HUMMER*. Motorbooks International, 1992.

Green Michael, and Stewart, Greg. *M2/M3 Bradley*. Concord Publications, 1990.

Greer, Don. *M1A1 Abrams in Action*. Squadron/Signal Publications, 1989.

Griess, Thomas E. *Campaign Atlas to the Second World War*. Avery Publishing Group, 1989.

Guderian, Heinz, translated by Christopher Duffy. *Achtung–Panzer!* Arms and Armour Press, 1992.

Gunston, Bill. *The Illustrated Encyclopedia of Aircraft Armament*. Orion Books, 1988.

Halberstadt, Hans. *NTC-A Primer of Modern Land Combat*. Presidio Press, 1989.

——*Army Aviation*. Presidio Press, 1990.

Hallion, Richard P. *Strike from the Sky-The History of Battlefield Air Attack 1911-1945*. Smithsonian Institution Press, 1989.

——*Storm over Iraq-Air Power and the Gulf War*. Smithsonian Institution Press, 1992.

Hammel, Eric. *Khe Sanh-Siege in the Clouds: An Oral History*. Crown Publishers, Inc., 1989.

Hansen, Chuck. *U.S. Nuclear Weapons-The Secret History*. Orion Books, 1988.

Hassler, Warren W. Jr. *Crisis at the Crossroads-The First Day at Gettysburg*. University of Alabama Press, 1970.

Harris, J. P., and Toase, F. N. (ed.). *Armoured Warfare*. B. T. Batsford Ltd., 1990.

Heinlein, Robert A. *Starship Troopers*. Ace Books, 1987.

Hilsman, Roger. *George Bush vs Saddam Hussein-Military Success/Political Failure?* Lyford Books, 1992.

Hudson, Heather E. *Communications Satellites: Their Development and Impact*. Free Press, 1990.

Hughes, B. P. *Open Fire: Artillery Tactics from Marlborough to Wellington*. Antony Bird Publications, 1983.

International Countermeasures Handbook. 12th ed. EW Communications, 1987.

The Iraqi Army: Organization and Tactics. Paladin Press, U.S. Army, 1991, p. 501.

Isby, David C. *Weapons and Tactics of the Soviet Army*. Jane's Publishing Company Ltd., 1981.

Isby, David C., and Kamps, Charles Jr. *Armies of NATO's Central Front*. Jane's Publishing Company Ltd., 1985.

Jane's Armour and Artillery Systems 1993-94. Jane's Information Group, 1993.

Keegan, John. *The Illustrated Face of Battle.* Viking Penguin Inc., 1989.

——*The Second World War.* Viking, 1989.

——*A History of Warfare.* Knopf, 1993.

Kelley, Orr. *King of the Killing Zone.* Berkley Books, 1989.

Kondo, Yoji (ed.). *Requiem.* Tom Doherty Associates, 1992.

Lee, R. G.; Garland-Collins,T. K.; Garnell, P.; Halsey, D. H. J.; Moss, G. M.; and Mowat, A.W. *Guided Weapons (Including Light, Unguided Anti-Tank Weapons).* Brassey's Battlefield Weapons Systems and Technology Series, Volume 8. Brassey's Publishers Ltd., 1983.

Lehman, John. *Making War.* Scribners, 1992.

Liddell-Hart, Basil Henry. *Strategy.* Praeger, 1967.

Luttwak, Edward, and Koehl, Stuart L. *The Dictionary of Modern War.* Harper Collins Publishers, 1991.

Macksey Kenneth. *Tank versus Tank-The Illustrated Story of Armored Battlefield Conflict in the Twentieth Century.* Crescent Books, 1991.

Macksey Kenneth, and Batchelor, John H. *Tank-A History of the Armoured Fighting Vehicle.* Ballantine Books, 1971.

McConnell, Malcolm. *Just Cause-The Real Story of America's High-tech Invasion of Panama.* St. Martin's Press, 1991.

McFarland, Stephen L., and Newton, Wesley P. *To Command the Sky.* Smithsonian Institution Press, 1991.

McWilliams, Barry. *This Ain't Hell...But You Can See It From Here!* Presidio Press, 1992.

Menninger, Bonar. *Mortal Error.* St. Martin's Press, 1992.

Mesko, Jim. *M2/M3 Bradley in Action.* Squadron/Signal Publications, 1992.

Morse, Stan (ed.). *Gulf Air War Debrief.* Aerospace Publishing, London, 1991.

Newhouse, John. *War and Peace in the Nuclear Age.* Knopf, 1989.

Nilsen, Robert. *South Korea Handbook.* Moon Publications, 1988.

Norman, Bruce. *Secret Warfare: The Battle of Codes and Cyphers.* David and Charles, 1989.

O'Ballance, Edgar. *No Victor, No Vanquished-The Yom Kippur War.* Presidio Press, 1978.

Pagonis, William G. *Moving Mountains: Lessons in Leadership and Logistics from the Gulf War*. Harvard Business School Press, 1992.

Peebles, Curtis. *Guardians-Strategic Reconnaissance Satellites*. Presidio, 1987.

Peeters, Willy. *AH-64A Attack Helicopter*. Verlinden Publications, 1991.

Peoples, Kenneth. *Bell AH-1 Cobra Variants*. Aerofax, Inc., 1988.

Perla, Peter P. *The Art of Wargaming*. Naval Institute Press, 1990.

Pfanz, Harry W. *Gettysburg: The Second Day*. University of North Carolina Press, 1987.

Phillips, Jeffrey, and Gregory, Robyn M. *America's First Team in the Gulf*. Taylor Publishing, 1992.

Popelka, Beverly A. (ed.). *Weapon Systems*. U.S. Army, 1992.

Pretty, Ronald T. *Jane's Weapon Systems, 1981-82*. Jane's Publishing Company, 1981

Prezelin, Bernard, and Baker, A. D. *Combat Fleets of the World, 1993*. Naval Institute Press, 1993.

Price, Alfred. *The History of U.S. Electronic Warfare*. Association of Old Crows, 1989.

Quarrie, Bruce. *Armoured Wargaming-A Detailed Guide to Model Tank Warfare*. Patrick Stephens Limited, 1988.

Richelson, Jeffrey T. *America's Secret Eyes in Space*. Harper & Row, 1990.

Rogers, Will, and Rogers, Sharon. *Storm Center: The USS Vincennes and Iran Air Flight 655*. Naval Institute Press, 1992.

Rommel, Erwin. *Infantry Attacks*. Presidio Press, 1990.

Santolli, Al. *Leading The Way-How Vietnam Veterans Rebuilt the U.S. Military*. Ballantine Books, 1993.

Schwarzkopf, H. Norman. *It Doesn't Take a Hero*. Bantam Books, 1992.

Shaara, Michael. *The Killer Angels*. Random House, 1974.

Simpkin, Richard E. *Antitank-An Airmechanized Response to Armored Threats in the 90s*. Brassey's Publishers Ltd., 1982.

Smallwood, William L. *Warthog-Flying the A-10 in the Gulf War*. Brassey's, 1993.

Smith, Peter C. *Close Air Support-An Illustrated History, 1914 to the Present*. Orion Books, 1990.

Smithfells, Colin J. *Metals Reference Book*. 4th ed., Volume III. Plenum Press, 1967.

Sorley, Lewis. *Thunderbolt–From the Battle of the Bulge to Vietnam and Beyond.* Simon & Schuster, 1992.

Starry, Donn. *Mounted Combat in Vietnam.* Department of the Army, 1978.

Stevenson, William. *90 Minutes at Entebbe.* Bantam, 1976.

Stewart, Greg. *National Training Center.* Concord Publications, 1992.

Summers, Harry G. *On Strategy II: A Critical Analysis of the Gulf War.* Dell, 1992.

Taylor, John W. R., and Munson, Kenneth. *Jane's All the World's Aircraft, 1984–85.* Jane's Publishing Company, 1984.

Terry, T. W. ; Jackson, S. R.; Ryley, C. E. S.; Jones, B. E.; and Wormell, P. J. H. *Fighting Vehicles, Land Warfare.* Brassey's New Battlefield Weapons Systems and Technology Series Volume 7. Brassey's (UK) Ltd., 1991.

Thompson, Julian. *No Picnic–3 Commando Brigade in the South Atlantic: 1982.* Hippocrene Books, 1985.

——*The Lifeblood of War: Logistics in Armed Conflict.* Brassey's (UK) Ltd., 1991.

Toffler, Alvin, and Toffler, Heidi. *War and Anti-War.* Little, Brown, 1993.

TRIMPACK GPS Receiver: Operation & Maintenance Guide. Trimble Navigation, 1990.

U.S. Army Field Manual 100-5: Blueprint for the AirLand Battle. Brassey's (U.S.), 1991.

U.S. News and World Report. *Triumph Without Victory.* Random House, 1992.

Vaux, Nick. *Take that Hill!–Royal Marines in the Falklands War.* Brassey's (U.S.) Inc., Maxwell Macmillian Pergamon Publishing Corp., 1986.

Von Clausewitz, Carl. *On War.* Penguin Classics, 1982.

Von Senger und Etterlin, Ferdinand. *Tanks of the World.* 7th edition. Bernard & Graefe Verlag, 1990.

Ward, Geoffrey C.; Burns, Ric; and Burns, Ken. *The Civil War.* Alfred A. Knopf, Inc., 1991.

Warden, John A. III. *The Air Campaign: Planning for Combat.* Brassey's (U.S.), 1989.

Watson, Bruce W.; George, Bruce M. P.; Tsouras, Peter; and Cyr, B. L. *Military Lessons of the Gulf War.* Greenhill Books and Presidio Press, 1991.

Weinberger, Caspar. *Fighting For Peace: Seven Critical Years in the Pentagon.* Warner Books, 1990.

White, B. T. *Tanks and other Armored Fighting Vehicles 1900-1918*. The Macmillian Company, 1970.

Winnefeld, James A., and Johnson, Dana J. *Joint Air Operations–Pursuit of Unity in Command and Control 1942–1991*. Naval Institute Press, 1993.

Winter, Frank H. *The First Golden Age of Rocketry*. Smithsonian Institution Press, 1990.

XM8 Armored Gun System. Program Briefing, FMC Corporation, 1993.

Zaloga, Steven J. *The M1 Abrams Battle Tank*. Osprey Publishing, London, 1985.

———*The M2 Bradley Infantry Fighting Vehicle*. Osprey Publishing, 1986.

———*Red Thrust - Attack on the Central Front, Soviet Tactics and Capabilities in the 1990s*. Presidio Press, 1989.

Zaloga, Steven J., and Green, Michael. *Tank Attack: A Primer of Modern Tank Warfare*, Motorbooks International Publishers & Wholesalers, 1991.

Zaloga, Steven J., and Sarson, Peter. *M1 Abrams Main Battle Tank: 1982-1992*. Osprey Publishing, 1993.

Zurick, Tim. *Army Dictionary and Desk Reference*. Stackpole Books, 1992.

Magazines

Air Force

Apache

Armada International

Armed Forces Journal

Armor- The Magazine of Mobile Warfare

Army

Army 1993-94 Green Book

Army Aviation

ATAC and the Armor/Anti-Armor Program

Aviation Week and Space Technology

Command

Field Artillery

Fighter Weapons Review

Flight International

Halting the Armoured Tide–JDW Survey

Helistop

International Defense Review

Jane's Defense Weekly

Jane's Intelligence Review

Jane's Soviet Intelligence Review

Military Technology (MILTECH)

Motor Trend

NATO's Fifteen/Sixteen Nations

Red Thrust Star

U.S. Naval Institute Proceedings

Warship International

World Airpower Journal

Brochures

"AH-1W Super Cobra." Bell Helicopter-Textron.

"AH-64A Apache–A Total System for Battle." McDonnell Douglas.

"Air-to-Air Stinger." General Dynamics Air Defense Systems.

"The Armed OH-58D Kiowa Warrior." Bell Helicopter-Textron.

"Army Tactical Missile System: Fact Sheet." Loral Vought Systems.

"AT4 Light Anti-Armor Weapon." BOFORS Weapon Systems.

"Avenger." Boeing.

"Bell OH-58D Kiowa Warrior." Bell Helicopter-Textron.

"Bradley A2-M2/M3 Fighting Vehicles." FMC Corporation.

"The Carl Gustaf System." BOFORS Weapon Systems.

"Combat Training and Simulation Systems." Loral Vought Systems.

"E-8C Joint STARS." Grumman.

"FDCV/CPV: Fire Direction Center Vehicle/Command Post Vehicle." BMY Corporation.

"Flexibility Sets the Pace at Combat Training Centers." Loral Vought Systems.

"Guided Weapon for T-80U and T-90E Tanks." Russian Brochure.

"Hellfire II Missile System." Martin Marietta Electronics Group.

"Hellfire–Ground-Launched Light Systems." Rockwell International, Tactical Systems Division.

"Hellfire Modular Missile System." Rockwell International, Tactical Systems Division.

"Hummer 25." AM General.

"HUMMER® M-988 Series: Specifications & Performance Data." AM General.

"Hydra-70." BEI Defense Systems, Fort Worth, TX.

"IVIS: Knowledge Is Power." General Dynamics Land Systems.

"Javelin Antitank Weapon System." Texas Instruments/Martin Marietta.

"Joint STARS." Grumman.

"LAW-M72 Light Anti-Armor Weapon." Talley Defense Systems.

"M1 Abrams Laser Rangefinder Thermal Imaging System." Hughes Aircraft Company, Electro-Optical Systems.

"M1 Evolution–M1A1." General Dynamics Land Systems.

"M1A1." General Dynamics Land Systems.

"M1A2–Fightability Defined." General Dynamics Land Systems.

"M1A2 Gunner's Primary Sight: Line-of-Sight Subsystem." Hughes Aircraft Company, Electro-Optical Systems.

"M1A2–Tomorrow's Solutions Today." U.S. Army.

"M9 Armored Combat Earthmover." BMY Corporation.

"M16A2 Rifle." Colt's Manufacturing Company, Hartford, CT.

"M16A3 Enhanced Family of Weapons." Colt's Manufacturing Company, Hartford, CT.

"M77 MLRS Rocket: Fact Sheet." Loral Vought Systems.

"M88A1 Recovery Vehicle." BMY Corporation.

"M88A1E1: The Improved Recovery Vehicle to Support M1 Series Tanks." BMY Corporation.

"M109A2 155mm Self-Propelled Howitzer." BMY Corporation.

"M113 Family of Vehicles: Modernizing for the Future." FMC Corporation.

"M113A3." FMC Corporation.

"M829A1 KE Tactical Cartridge." Olin Ordnance, St. Petersburg, FL.

"M830 HEAT Tactical Cartridge." Olin Ordnance, St. Petersburg, FL.

"Meet an American Legend: Hummer®." AM General.

"MILES Air-to-Ground Engagement System II." Loral Vought Systems.

"MLRS." Loral Vought Systems.

"Multiple Launch Rocket System: Fact Sheet." Loral Vought Systems.

"National Training Center." U.S. Army.

"Non-Line-of-Sight Combined Arms (NLOS-CA)." Boeing.

"120mm Tank Ammunition." AllianTechSystems Precision Armament Systems.

"Oshkosh M1070 Heavy Equipment Transporter Specifications." Oshkosh
    Truck Corporation.

"Oshkosh PLS: Palletized Load System Specifications." Oshkosh Truck
    Corporation.

"Precision Lightweight GPS Receiver (PLGR)." Trimble Navigation.

"RAH-66 Comanche: Now...More than Ever." Boeing-Sikorsky.

"Scout M GPS Handheld Tactical SPS Receiver." Trimble Navigation.

"Shield – T-72S Rocket Tank." Russian Brochure.

"Stinger Family of Weapon Systems." Hughes Missile Systems. Pomona, CA.

"T-80U, PROMEXPORT." Russian Brochure.

"Tank Ammunition." AllianTechSystems, Brooklyn Park, MN.

"TOW 2 / TOW 2A." Hughes Aircraft Company.

"TOW 2B–A Fly-Over Shoot-Down TOW." Hughes Aircraft Company.

"TRADOC: Where Tomorrow's Victories Begin." U.S. Army.

"TRIMPACK, AN/PSN-10(v) Small, Lightweight GPS Receiver." Trimble
    Navigation.

"TRIMPACK Quick Reference Guide, Revision C." Trimble Navigation.

"UH-60L Blackhawk." Sikorsky.

"U.S. Army AH-64A Apache." McDonnell Douglas.

"U.S. Army/Rockwell Hellfire Modular Missile." Rockwell International,
    Tactical Systems Division.

"XM8 Armored Gun System." FMC Ground Systems Division.

Monographs

Frizzell, D., and Bowers, R. (ed.). "Air Power and the 1972 Spring Invasion." USAF Southeast Asia Monograph Series.

Gawrych, George W. "Key to the Sinai: The Battles for Abu Ageila in the 1956 and 1967 Arab-Israeli Wars." U.S. Army Command and General Staff College, 1990.

Kamiya, Jason K. "A History of the 24 Mechanized Infantry Division Combat Team during Operation Desert Storm." U.S. Army, 1991.

Mesko, Jim. "M60 Patton in Action." Squadron/Signal Publications, 1986.

Netherland, Scott F. "Use of the Global Positioning System by U.S. Forces during the Gulf War." German Army Conference on GPS, 1991.

Romjue, John L. "From Active Defense to AirLand Battle: The Development of Army Doctrine 1973-1982." U.S. Army, TRADOC Historical Monograph Series, 1984.

Pamphlets

"AFV." Profile Publications Limited, 1993.

"AH-64A Apache Anti-Armor Helicopter System Description." McDonnell Douglas Helicopter Company, 1986.

"Apache AH-64A." McDonnell Douglas Helicopter Company, 1993.

"Bellona Military Vehicle Prints." Bellona Publications Ltd., 1993.

"Bradley Derivative Vehicle Systems." FMC Corporation, 1993.

"Comanche: RAH-66." Boeing-Sikorsky, 1992.

"Defense Systems Group: Program Status." FMC Corporation, 1993.

"The Desert Jayhawk: Operation Desert Shield/Storm." U.S. Army, 1991.

"Desert Storm Conference Report." U.S. Army, 1992.

"General Dynamics 1992 Shareholder Report, General Dynamics." Corporate Headquarters, Falls Church, VA, 1992.

"German Tanks and Armoured Vehicles 1914-1945." Ian Allen Ltd., 1966.

"GTA-17-2-13 Armored Vehicle Recognition." U.S. Army, 1984.

"History, Customs, and Traditions of the 3d Armored Cavalry Regiment." U.S. Army, 1992.

"Key Weapons and Equipment Guide: Warsaw Pact Armies." U.S. Army, 1974.

"Leadership and Command on the Battlefield: Battalion and Company." U.S. Army, 1993.

"Program and Operational Highlights of the Armed OH-58D Kiowa." American Helicopter Society, 1990.

"Sikorsky H-60 Product Line." Sikorsky, 1992.

"State of America's Army on Its 218th Birthday." U.S. Army, 1993.

"Team Apache Modernization: Lifting the Fog of War." McDonnell Douglas Helicopter Company, 1993.µ

"Thinking About the Army's Future: Continuity, Change and Growth." U.S. Army, 1993.

"Trimble Navigation Annual Report." Trimble Navigation, 1991.

"U.S. Army Advanced Concepts and Technology Program." U.S. Army, 1993.

Briefing Notes

"Apache Program Status." McDonnell Douglas Helicopter Company.

"Brave Rifles 101, Third Armored Cavalry Regiment." U.S. Army.

"Comanche: RAH-66." Boeing-Sikorsky.

"MLRS Overview, Schaefer, Walter." FMC Corporation.

"XM93 & XM93E1." General Dynamics Land Systems Division.

Games

"MBT." Day, James M. (design and research). Avalon Hill, 1989.

"Phase Line Smash." Chadwick, Frank. GDW, 1992.

Maps

East Africa, 1:2,700,000. Karto+Grafik, 1989.

Soul (Seoul) Korea, 1:1,000,000. Army Map Service, 1964.

Virginia, 1:50,000. Alexandria. Defense Mapping Agency, V734X55611.

Washington West Quadrangle; District of Columbia, Maryland, Virginia, 1:24,000. U.S. Geological Survey, 1983.

Videotapes

ABC News, Mr. Donaldson & CG. Battle Labs, TRADOC Command BFG.

A More Cunning Fox. General Dynamics Land Systems.

Apache Owns the Night. McDonnell Douglas Helicopter Company.

Armed and Dangerous. McDonnell Douglas Helicopter Company.

Armored Gun System-Executive Summary. FMC Corporation.

Armored Gun System-Progress Review I. FMC Corporation.

Armored Gun System-Progress Review II. FMC Corporation.

Army Chief of Staff General Gordon Sullivan Visits the Boeing-Sikorsky Comanche Team. Sikorsky Aircraft.

AUSA. AM General.

Bell Helicopters in the Gulf War.

The Big Picture. FMC Corporation.

BMY Combat Systems King of Battle.

BMY Combat Systems Meeting the Challenge.

Bradley Fighting Vehicle Performance in SWA: A Conversation with Colonel Douglas Staar, Commander 3rd Cavalry. FMC Corporation.

Bradley...the Soldiers' Vehicle. FMC Corporation.

C2V Interior & Exterior Views. FMC Corporation.

The Civil War. PBS Home Video.

CYPHER Free Flight Demonstration Tape #2. Sikorsky Aircraft.

Desert Storm Chronicles. U.S. Army.

First Flight AH-64D. McDonnell Douglas Helicopter Company.

Forged in Fire. McDonnell Douglas Helicopter Company.

Fox NBCRS. General Dynamics Land Systems.

General Franks, 73 Easting. U.S. Army.

Hellfire-The Difference AUSA '92. Rockwell International.

HET M-1070. Oshkosh Truck Corporation.

Hummer: The Inside Story. AM General.

LAM: The Update.

Longbow AAAA '92. McDonnell Douglas Helicopter Company.

M1A1 Technical Characteristics. General Dynamics Land Systems.

M1A2 Fightability Defined. General Dynamics Land Systems.

M1A2-The Decisive Edge. General Dynamics Land Systems.

M113 Family of Vehicles Modernizing for the Future.

M113 FOV Acceleration Comparison Tests. FMC Corporation.

M113A3 Universal Carrier. FMC Corporation.

M992A1 FAASV-Attacking the Munitions Challenge. BMY Corp.

Mad Dogs in Saudi. U.S. Army.

MGEN McCaffrey Troop Talks. U.S. Army.

MLRS-In the Storm. Loral Vought Systems.

MLRS-The Making of a Winner. Loral Vought Systems.

MLRS-Total Victory. FMC Corporation.

Modernized Apache Operational Capabilities. McDonnell Douglas Helicopter Company.

NATO/UN Operations at CMTC. U.S. Army.

NBC Nightly News "What Works" Hummer. AM General.

Night of the Apache. McDonnell Douglas Helicopter Company.

Partners in Success. General Dynamics Land Systems.

PLS: The Army's Total Distribution System. Oshkosh Truck Corporation.

RAH-66 Comanche PPR 2nd Gen FLIR Update 3/92 Fantail (music video). Sikorsky Aircraft.

Reconnaissance: The Key to Victory ETV. Fort Rucker, U.S. Army.

Simulation Insights. U.S. Army.

Simulation Insights and the Reconstruction of the Battle of 73 Easting.

U.S. Army Simulation. Sikorsky Aircraft.

War in the Gulf. Video Ordnance.

Warrior on the Move. McDonnell Douglas Helicopter Company.

Wings of Apache. McDonnell Douglas Helicopter Company.

Wings Over the Gulf. Discovery Communications, Inc.